Watershed Development Projects In India

An Evaluation

John Kerr, in collaboration with Ganesh Pangare and Vasudha Lokur Pangare

RESEARCH REPORT 127

INTERNATIONAL FOOD POLICY RESEARCH INSTITUTE
WASHINGTON, D.C.

Printed in the United States of America

International Food Policy Research Institute
2033 K Street NW
Washington, DC 20006-1002
+1-202-862-5600
www.ifpri.org

Library of Congress Cataloging-in-Publication Data

Kerr, J. M.
Watershed development projects in India : an evaluation / John Kerr, in collaboration with Ganesh Pangare and Vasudha Lokur Pangare.
p. cm.— (Research report ; 127)
Includes bibliographical references.
ISBN 0-89629-129-4 (alk. paper)
1. Dry farming—India—Maharashtra. 2. Dry farming—India—Andhra Pradesh. 3. Agricultural productivity—India—Maharashtra. 4. Agricultural productivity—India—Andhra Pradesh. 5. Water resources development—India—Maharashtra. 6. Water resources development—India—Andhra Pradesh. I. Pangare, Ganesh. II. Pangare, Vasudha. III. Title. IV. Research report (International Food Policy Research Institute) ; 127
SB110 .K47 2002
631.5'86'0954—dc21 2002007983

Contents

Tables

Figures

Foreword

In recent years India has looked to watershed development as a way to realize its hopes for agricultural development in rainfed, semi-arid areas. These areas were bypassed by the Green Revolution and have experienced little or no growth in agricultural production for several decades. By capturing scarce water resources and improving the management of soil and vegetation, watershed development has the potential to create conditions conducive to higher agricultural productivity, while conserving natural resources.

This case study of watershed projects in Andhra Pradesh and Maharashtra, India, fits squarely into IFPRI's work on development strategies for less-favored lands—areas where agroclimatic conditions are difficult or infrastructure and support services have been neglected. It offers important insights for IFPRI's research in other parts of the world, including work under way in the East African highlands, the hillsides of Central America, and the West Asia and North Africa regions.

While much has been written about watershed development, there have been few efforts to systematically evaluate it. By doing so, John Kerr and his colleagues Ganesh Pangare and Vasudha Lokur Pangare contribute immensely to our understanding of the promise and challenges of watershed development. Their report is critically important for at least three reasons.

First, much of the hype surrounding Indian watershed development draws on a small number of extraordinary success stories that may be difficult or impossible to replicate on a large scale. Studying a large number of randomly selected projects makes it possible to see beyond these unusual successes.

Second, watershed development has evolved from a purely technical, externally imposed intervention in the 1980s to a more participatory exercise in which local people help design and implement management plans. This study examines a broad range of watershed approaches and compares their performance in improving natural resource management and raising agricultural production.

And third, watershed development is popular among development planners and agricultural scientists because it promises a win-win solution in which natural resource conservation and agricultural productivity are complementary. It is often assumed that watershed development will also have favorable implications for poverty alleviation, but it may actually impose hardships. This study examines this question in detail.

The authors conclude that while most of the projects they surveyed have had relatively little impact, those that take a more participatory approach and are managed by NGOs have in fact performed better in conserving natural resources and raising agricultural productivity. But such success may come at the expense of the poorest people in watershed areas because improving the management of a watershed usually requires restricting access to the natural resource base on which they depend. Many watershed development projects do not work because those whose interests are harmed refuse to go along with the effort. The authors argue that for watershed development to succeed on a large scale, projects must find a way for all affected parties to share in the net gains generated.

Joachim von Braun
Director General, IFPRI

Acknowledgments

The authors wish to thank Peter Hazell, Derek Byerlee, John Pender, Anna Hazare, P. J. George, Shashi Kolavalli, John Farrington, G. B. Singh, Dayanatha Jha, Shahidur Rashid, two anonymous reviewers, and numerous investigators and respondents who contributed to the study. The National Centre for Agricultural Economics and Policy Analysis (NCAP) provided valuable assistance and generous use of its facilities.

Summary

The Green Revolution that transformed agriculture elsewhere in India had little impact on rainfed agriculture in the semi-arid tropical regions, where agricultural productivity is low, natural resources are degraded, and the people are poor. In the 1980s and 1990s, agricultural scientists and planners aimed to promote rainfed agriculture through watershed development. A watershed is an area from which all water drains to a common point, making it an attractive unit for technical efforts to manage water and soil resources for production and conservation. Watershed management is complicated, however, by the fact that watersheds rarely correspond to human-defined boundaries. Also, watershed projects often distribute costs and benefits unevenly, with costs incurred disproportionately upstream, typically among the poorer residents, and benefits realized disproportionately downstream, where irrigation is concentrated and the wealthiest farmers own most of the land.

Watershed project strategies range from those that are more technocratic to those that pay more attention to social organization. By the late 1990s annual expenditure on watershed development in India exceeded US$500 million, but relatively little information was available on the success of different project approaches.

This study addresses three main research questions: (1) What projects are most successful in promoting the objectives of raising agricultural productivity, improving natural resource management, and reducing poverty? (2) What approaches enable projects to succeed? (3) What nonproject factors contribute to achieving these objectives?

A detailed survey of Maharashtra and Andhra Pradesh states, conducted as part of this research in 1997, covered 86 villages, some included in watershed projects and some not. The projects covered operated under India's Ministry of Agriculture and Ministry of Rural Development, under various nongovernment organizations (NGOs), and under NGO and government of Maharashtra collaboration. The government projects focused largely on technical improvements; the NGO projects focused more on social organization; and the collaborative projects tried to draw on the strengths of both approaches. (The survey was conducted at a time when the Ministry of Rural Development's participatory guidelines were too new to be used to evaluate the projects, but the analysis is still able to provide insights relevant to these projects because the new guidelines were modeled on NGO approaches already in use.)

The analysis compares conditions in the study villages before and after the projects were implemented. Quantitative analysis at the village level examines performance indicators such as changes in access to irrigation water, employment opportunities, soil erosion and conservation on uncultivated lands and drainage lines, and availability of various products such as fodder from the common (government revenue) lands. At the plot level, performance indicators include changes in cropping intensity, yields, soil erosion on cultivated lands, farmers' land improvement investments, and annual net returns to cultivation. An instrumental variables econometric approach is used to correct for endogenous program placement. This analysis is supplemented by qualitative information about the effects of the projects on different interest groups in the villages such as farmers with and without irrigation, landless people, shepherds, and women.

In both Maharashtra and Andhra Pradesh, the participatory projects performed better than their technocratic, top-down counterparts. However, participation combined with sound technical input performed best of all. For example, while all projects reduced soil erosion on uncultivated lands reasonably well, the NGO and NGO/government collaborative projects had particularly good records in this regard, probably because they effectively introduced social institutions to limit exploitation of uncultivated lands. Although definitive hydrological data were not available, farmers in villages in NGO and NGO/government project villages frequently perceived that the projects' water-harvesting efforts increased the availability of water for irrigation and their net returns to rainfed farming were higher.

The better performance of the more participatory projects seems to be related to the complex, often site-specific livelihood systems prevalent in the study areas. These conditions call for a flexible approach and responsiveness to diverse, often unexpected situations. Blueprint approaches pursued by technocratic, hierarchical organizations are poorly suited to such conditions. The NGO and NGO/government collaborative projects devoted time and resources to organizing communities to establish locally acceptable social arrangements for watershed interventions. They selected villages where people had demonstrated the ability and willingness to work collectively to solve common problems. Such screening makes sense where budgets are limited and the list of failed projects is long.

Problems remain regarding equity. By their nature area development programs offer benefits primarily to landowners, with landless people benefiting indirectly, either through peripheral program activities or trickle-down effects. In fact, watershed projects can actually make women and landless people worse off by restricting their access to resources that contribute to their livelihoods. Even some of the more participatory projects have found it difficult to ensure that benefits reach all of the intended population: respondents perceived that project benefits rose with landholding size, and landless and near-landless people were most likely to report negative effects from projects. The very best projects help the poorest and politically weakest community members negotiate with their neighbors to ensure that everyone benefits, but this remains an area where nearly all projects need to improve.

The government's new, participatory guidelines, designed along the lines of the NGO and NGO/government collaborative projects, should have favorable results. But one important caveat must be considered: the successful projects—few in number—may have enjoyed special attention that cannot be replicated on a large scale. In rainfed areas of Maharashtra's Pune and Ahmednagar districts, for example, the innovative projects operated in only about 40 out of a 1,000 villages, even though they were more concentrated there than in the rest of India. Scaling up will stretch the supply of good NGOs and will reduce the attention that any given village can receive from the highest quality people. Once the political spotlight on a single village is removed, there is less urgency to ensure project success.

The installation of soil and water conservation measures on both private and common land were subsidized nearly 100 percent, largely to generate employment. But where subsidies are high, long-term maintenance of the asset is often low. This problem arises because, to gain employment, people will accept conservation measures they do not really want. To improve soil conservation without dismantling employment benefits, project designers will have to find more sophisticated ways to ensure that people select the measures they really want. Some projects are experimenting with removal of subsidies for private soil conservation investments.

This research uncovered little evidence either to support or deny the hypothesis that good infrastructure and other favorable economic conditions promote better conservation performance. Infrastructure conditions changed little over time, so there was insufficient variation in the sample to detect patterns.

Finally, the lack of existing monitoring and evaluation data for watershed projects was an important constraint. Monitoring and evaluation are essential for assessing new, innovative approaches to watershed development.

CHAPTER 1

Introduction

Rainfed agriculture in India's semi-arid tropics is characterized by low productivity, degraded natural resources, and widespread poverty. Most of the hundreds of millions of people living in the Indian semi-arid tropics depend on agriculture and natural resource management for their livelihoods, so development planners are eager to implement productive, environmentally sustainable land and water management systems.

Watershed development projects are designed to harmonize the use of water, soil, forest, and pasture resources in a way that conserves these resources while raising agricultural productivity, both by conserving moisture in the ground and increasing irrigation through tank- and aquifer-based water harvesting. Watershed projects have become widespread in rainfed areas in recent years, with a current annual budget from all sources that exceeds US$500 million (Farrington, Turton, and James 1999). This report examines the experience of watershed projects in Andhra Pradesh and Maharashtra.[1]

The literature on watershed development in India is growing rapidly, but most of it is confined to qualitative descriptions of success stories. Some of these contain excellent insights into the social processes that contribute to successful watershed development, but there is little frank discussion of less successful projects. The few quantitative studies available tend to be based on a small number of heavily supervised projects, with no information about long-term effects. Benefits after the first year or two were typically assumed, and, not surprisingly, cost-benefit findings were almost always favorable. At the same time, the vast majority of projects were never evaluated, and there were good reasons to suspect that most of them had little impact (Kerr and Sanghi 1992).

With this background, the current research was commissioned jointly by the World Bank and ICAR to analyze the determinants of agricultural productivity, natural resource management, and poverty alleviation under a wide range of watershed projects. The research is mainly quantitative but also incorporates qualitative data, explicitly examining the effects of factors unrelated to the projects such as infrastructure, access to markets, social institutions in the villages, agroecological conditions, and so forth. This broad framework not only controls for the effects of these factors but also enables identification of other policy-relevant determinants of improved natural resource management and economic development. The approaches taken by different projects are discussed in order to understand the essential elements of successful projects and make recommendations for the future. To summarize, the study addresses

[1]This study was originally conducted under the Indian Rainfed Agricultural Research and Development Project, jointly sponsored by the World Bank and the Indian Council of Agricultural Research (ICAR).

three related questions. (1) Which projects performed the best? (2) What approaches enabled them to succeed? And (3) What additional characteristics of particular villages contributed to achieving the objectives of improved natural resource management, higher agricultural productivity, and reduced poverty?

Research Issues

Two main hypotheses guided this research. First, watershed projects cannot succeed without full participation of project beneficiaries and careful attention to social organization. This is because the costs and benefits of watershed interventions are location-specific and unevenly distributed among the people affected. Second, economic conditions and access to infrastructure may have as great an impact as a watershed project in determining the outcomes that projects seek to achieve because such factors determine the incentives for people to manage and protect natural resources and invest in increased agricultural productivity.

Watershed Management as a Social Organization Problem

A watershed (or catchment) is a geographic area that drains to a common point, which makes it an attractive unit for technical efforts to conserve soil and maximize the utilization of surface and subsurface water for crop production. A watershed is also an area with administrative and property boundaries, lands that fall under different property regimes, and farmers whose actions may affect each other's interests. Boundaries defined by humans, however, normally do not match biophysical ones. In watershed management projects, mechanical or vegetative structures are installed across gullies and rills and along contour lines, and areas are earmarked for particular land use based on their land capability classification. Cultivable areas are put under crops according to strict principles of contour-based cultivation. Erosion-prone, less favorable lands are put under perennial vegetation. This approach aims to optimize moisture retention and reduce soil erosion, thus maximizing productivity and minimizing land degradation. Improved moisture management increases the productivity of improved seeds and fertilizer, so conservation and productivity-enhancing measures are complementary.

Excess surface runoff water is harvested in irrigation or percolation tanks while subsurface drainage recharges groundwater aquifers, so conservation measures in the upper watershed have a positive impact on productivity in the lower watershed. Reducing erosion in the upper reaches also helps to reduce sedimentation of irrigation tanks (ponds) in the lower reaches. The watershed approach enables planners to harmonize the use of soil, water, and vegetation in a way that conserves these resources and maximizes their productivity. This systems-based approach is what distinguishes watershed management from earlier plot-based approaches to soil and water management.

Socioeconomic relationships among people in a watershed can complicate efforts to introduce seemingly straightforward technical improvements. This is because a watershed encompasses many decisionmakers who are affected unequally by development. When a watershed project is introduced, often the bulk of the work is done in the upper reaches, while the benefits accrue primarily to those in the lower reaches. For example, revegetating the upper reaches involves imposing a ban on both grazing of animals and felling of trees so that plants can become established. As a result, the people who use the upper watershed—typically relatively poor people with little or no land—bear the brunt of the costs of watershed development, which mainly benefits wealthier farmers in the lower watershed. Those who are made worse off by a watershed project can undermine its efforts if they refuse to go along with it. Herders, for example, might refuse to abide by grazing bans and trespass on the common lands if they can. In general, watershed technologies are likely to fail if they divide benefits unevenly but require near-universal cooperation to make them work. In this case, equity becomes a prerequisite to efficiency (Kerr and Sanghi 1992).

While early watershed projects failed to recognize the socioeconomic dimensions of watershed development, this has changed significantly in the last decade. In recent years, there has been a growing appreciation of the need to organize communities to work collectively, to ensure that everyone takes an interest in the work that is done and benefits from the project. In the 1990s, every project was designed to include the "participation" of local people; however, each project defined participation differently. For government programs, typically it meant making the effort to convince people of the soundness of an approach that was essentially predesigned without any input from those who would be affected. Some projects took people's involvement a step further: local committees were established to mobilize laborers to move earth and plant vegetation and to facilitate communication within the village to improve the management of common lands. At the extreme, many new projects operate under the assumption that local people know best how to care for their land and simply need outside assistance to help them organize and gain access to resources, including funds and social services.

Approaches to participation are discussed in detail in Chapter 2, and implications of alternate approaches for project outcomes are revealed by the analytical findings presented in Chapters 6–8. Based on these findings and various observations from the field, recommendations for how projects should pursue participation in the future are presented in Chapter 9.

How Economic Forces Determine Project Outcomes

Improving agricultural production, natural resource management, and human welfare depends on economic factors beyond the control of a watershed project. Throughout the world, both today and historically, it is easy to find areas with a broad range of performance in agricultural growth, natural resource management, and poverty alleviation. For example, areas abound in India with stagnant agricultural production, low real incomes, and environmental degradation. But both the literature and folk wisdom are full of examples of places in India where villagers manage their natural resources particularly well and the local economy is unusually vibrant. What determines why some areas are more productive than others?

A theory called "induced innovation" helps explain the conditions under which agricultural development will take place along paths that degrade or conserve natural resources. Induced innovation theory holds that, over time, technological innovations and institutional changes take place to economize on scarce resources and utilize abundant ones (Hayami and Ruttan 1984). The theory helps explain why traditional farming systems have evolved differently in different places. For example, in sparsely populated areas bush-fallow was the traditional farming system. Under this system forest land is cleared and farmed for a few years; then it is left fallow for 20 to 30 years to restore nutrients. In land-scarce areas such as the intensive rice-growing areas of Southeast Asia, however, elaborate terraces, irrigation systems, and nutrient management systems enable farmers to cultivate the land continuously without degradation. In the widely cited case of Machakos,

Kenya, rising population density, good access to markets, and off-farm income created incentives and provided resources to raise productivity and conserve natural resources (Tiffen, Mortimore, and Gichuki 1994).

In India, farmer-led agricultural intensification is also widespread. In semi-arid areas the most obvious example is that of private irrigation investments, which are typically accompanied by land leveling and application of substantial organic matter and commercial inputs. On rainfed lands the successes are less dramatic, but evidence shows that private tree planting has grown steadily in recent years (Chambers, Shah, and Saxena 1989), and that many farmers invest in indigenous soil and water conservation measures independent of special project efforts (Kerr and Sanghi 1992). Likewise, some villages have designed social institutions for managing common property resources in ways that raise their productivity and protect against long-term resource degradation (Wade 1988).

Several exceptional case studies of successful watershed development have been well publicized in India, but the common perception is that they remain just that: exceptional. Success is often attributed to the efforts of a charismatic leader or some other set of social conditions that would be difficult or impossible to replicate on a wide scale. There is undoubtedly a great deal of truth in this perception, but to date there has been little systematic effort to examine the extent to which policy-relevant factors have played a role in causing some areas to be characterized by better resource management and higher agricultural production than others. Leaving aside unusual success stories like Ralegan Siddhi (Hazare, Pangare, and Lokur 1996) and Sukhomajri (Chopra, Kadekodi, and Murty 1990; Patel-Weynand 1997), are there village-level or regional differences in natural resource conditions, agricultural productivity, and household incomes that can be explained by induced innovation theory?

From the induced innovation perspective, assessing the performance of watershed development projects requires examining the effects of such factors as market access, population density, and the economic policy environment. Induced innovation theory suggests that if market access is favorable and population density is high, people will be more receptive to projects seeking to conserve soil resources and intensify agricultural production. In fact, even in the absence of a special project, the economic environment may be sufficient to induce farmers to adopt resource-conserving, productivity-enhancing technologies. On the other hand, even a well-designed watershed development project might be unable to achieve long-term success if enabling conditions are lacking. In such cases, farmers would not be sufficiently motivated to adopt and maintain practices that promote sustainable agricultural intensification.

Analytical Approach

This study examines performance in improving agricultural productivity, natural resource management, and human welfare. Data on performance indicators come from a 1997 survey conducted by the International Food Policy Research Institute (IFPRI) in collaboration with the National Centre for Agricultural Economics and Policy (NCAP), an Indian government research institute. The survey of 86 villages in Maharashtra and Andhra Pradesh (Figure 1.1) reflects a variety of project approaches and includes data on villages with no projects. Quantitative data collected at the village, plot, and household level provide the basis for econometric analysis of the conditions that determine change before and after the project. Further qualitative information on the impact of projects on people from

Figure 1.1 — Map of Maharashtra and Andhra Pradesh States

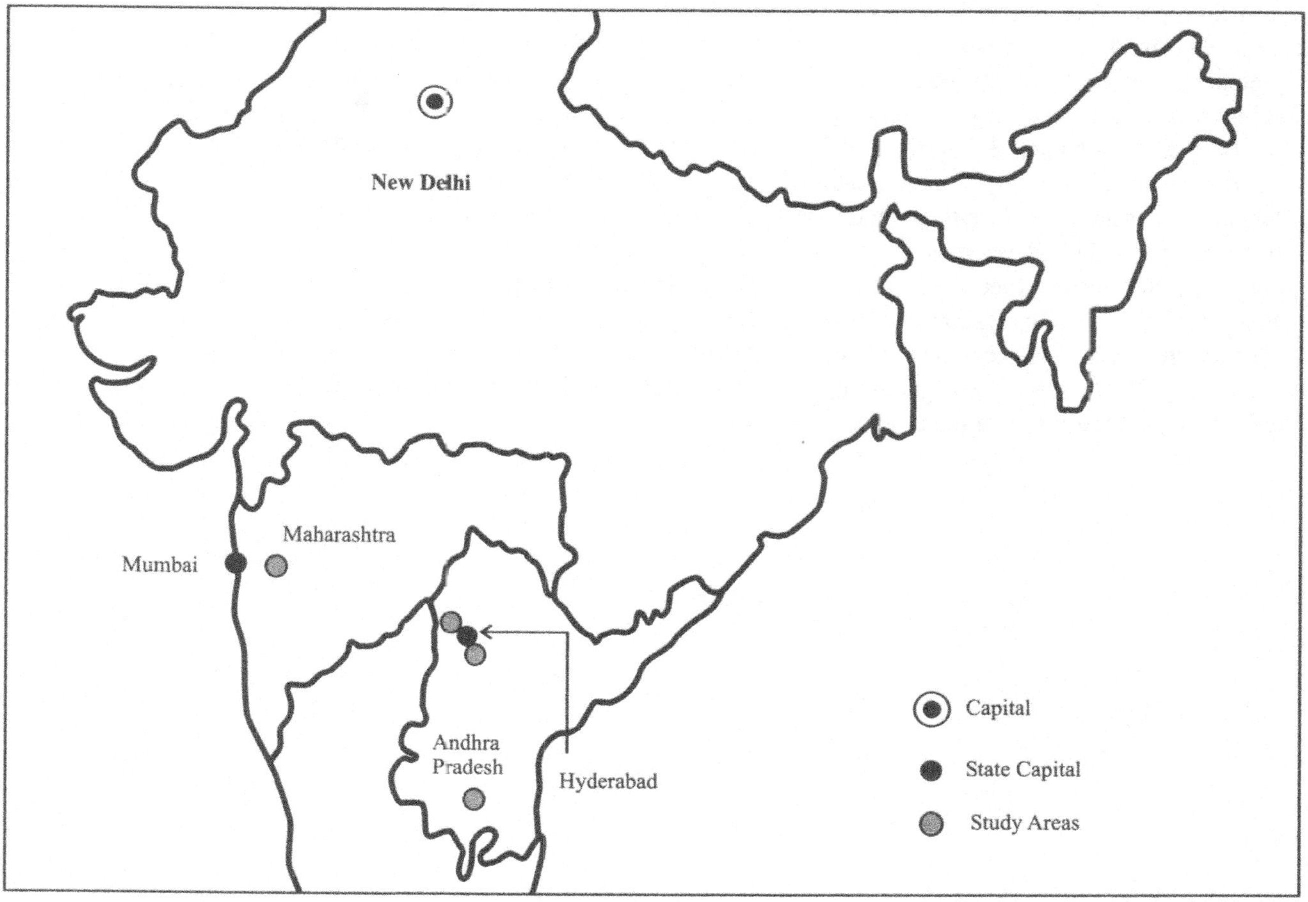

various interest groups, such as farmers with and without irrigation and livestock herders, was obtained through open-ended discussions.

This research was originally designed to examine only completed projects where the staff had withdrawn. However, despite the large literature on watershed development in India, the number of projects in which work had actually been completed was quite small, so the intended approach was not feasible. Instead, the survey covered mainly well-established projects, only a few of which had been completed.

One important consideration is that projects may select villages with certain characteristics, or villages may self-select for participation in a given project. Under these circumstances there may be systematic but unobserved differences among villages in different projects. If so, when attempting to assess projects' impacts, it may be difficult to distinguish between the effects of project interventions and those of preexisting conditions. To adjust the econometric analysis to account for this problem, an instrumental variables approach is used (see Chapter 4).

Organization of the Report

Chapter 1 introduces the problem and presents a conceptual framework. Chapter 2 describes the broad approaches to watershed development in Maharashtra and Andhra Pradesh and introduces the specific projects operating there. Chapter 3 describes the data on which this study is based, and Chapter 4 presents the analytical model. Characteristics of the villages in which each project operates are analyzed in Chapter 5, while Chapters 6 to 8 analyze project performance in terms of achievement of the various objectives related to agricultural productivity, natural resource management, and poverty alleviation. These include the work conducted by watershed projects on protecting and developing nonarable land, recharging groundwater, improving the management of agricultural land, and raising agricultural production. These chapters also analyze the role of nonproject factors such as infrastructure development on outcomes of interest, and they examine watershed project activities in relation to villagers' development priorities. Chapter 9 concludes with policy implications and recommendations. A glossary defines the terms and lists the acronyms used in the report.

CHAPTER 2

Approaches to Watershed Development in Maharashtra and Andhra Pradesh

This chapter describes the agroclimatic characteristics of the study region and characterizes two alternative technical approaches that have been used in watershed development. It describes each of the projects covered, focusing on their guiding principles and the relative emphasis on social organization compared with technical assistance. It considers how each project selects the sites where it works, the amount of money it will invest, and its policies regarding cost-sharing with intended beneficiaries.

Agroclimatic Conditions in Maharashtra and Andhra Pradesh

Maharashtra and Andhra Pradesh both have highly diverse agroclimates. In Maharashtra, a narrow coastal plain separates the Arabian Sea from the Western Ghat Mountains. On the eastern side of the mountains, the majority of the state is spanned by the large Deccan Plateau, which covers much of south-central India. Rainfall is very high in the coastal mountains, but the western part of the Deccan Plateau (in the rain shadow of the Ghats) is very dry. The wettest district of the Western Ghats receives an annual average rainfall of more than 4,000 millimeters, while the driest areas of the rain shadow zone (only about 150 kilometers to the east) receive about 500 millimeters. The topography of this transitional zone from wet to dry is a series of tablelands—flat plateaus that drop sharply to plains below. Conditions for rainfed agriculture in the driest zones are difficult, and this is where watershed projects are most often concentrated. Moving toward eastern Maharashtra, average annual rainfall rises gradually to more than 1,000 millimeters, making conditions for rainfed agriculture quite favorable.

Andhra Pradesh is similarly diverse. The long coastal plain along the Bay of Bengal receives more than 1,000 millimeters of average annual rainfall, and much of it is irrigated by the major canal systems of the Krishna and Godavari rivers. Moving west from the coast and over the Eastern Ghats (which are much smaller than the western Ghats), inland areas on the Deccan Plateau are divided into the Rayalseema and Telengana regions. Rayalseema is the southernmost part of the state; it is highly drought prone with average annual rainfall as low as 500 millimeters in some areas. South Telengana (which is south of Hyderabad) is also drought prone, though not to the same extent, with average annual rainfall in the 600–700 millimeters range. Both Rayalseema and south Telengana vary in their topography, with small hills and valleys that are suitable for traditional irrigation tanks that capture runoff from rainfall for lowland irrigation. The predominantly red soils of these regions also favor tank irrigation. North Telengana (which is north of Hyderabad), on the other hand, is flatter, has black soils, and receives about 800–1,000 millimeters of average annual rainfall. Conditions are much better for rainfed agriculture, comparable to the conditions across the state border in eastern Maharashtra.

Thus, rainfed agriculture in both states varies between areas of high and low potential. This heterogeneity has important implications for the approaches to watershed development.

How Different Projects Approach Raising Agricultural Productivity

There are fundamental differences between watershed projects that focus on developing rainfed agriculture and those that focus on increasing access to irrigation.

Projects that Focus on Increasing Irrigation

In western Maharashtra, the scarcity of water and favorable topography make water harvesting a high priority and the focus of most projects. Where plateaus slope down to the plains, there are many opportunities to capture water behind small dams for irrigation in the flat lands below. Soils in these areas are relatively porous and favor percolation of harvested water into groundwater aquifers; water must be pumped for irrigation use. By contrast, in Telengana and Rayalseema regions of interior Andhra Pradesh, irrigation tanks store water on the surface for irrigation by gravity.

Agricultural engineering to build and protect water-harvesting structures is the key feature of most watershed projects in western Maharashtra. The structures include check dams—small stone or earthen structures—in drainage lines and continuous contour trenches in the uncultivated catchment areas. Since almost all the structures are built on nonarable lands to which all village inhabitants have access, the projects also promote collective action to protect vegetation in the catchment area. This reduces erosion and silting that would reduce the storage capacity of water-harvesting structures.

In these projects there is relatively little focus on plot-level management. Once irrigation is in place, farmers have sufficient knowledge and incentive to manage a plot and improve its productivity. Rainfed agriculture is a low priority where projects are successful in increasing irrigated area. For example, Shri Anna Hazare, who is known as the "father of watershed development" as a result of the well-known success of his efforts in the village of Ralegan Siddhi, explained that watershed efforts there focus exclusively on increasing irrigation and protecting nonarable lands. Virtually no attention is paid to developing rainfed agriculture.[2] This approach has proven highly successful in Ralegan Siddhi, where irrigated area has gone from virtually zero to about 70 percent of the cultivated land over the last 25 years. Average annual rainfall is barely 500 millimeters, so conditions are not favorable for rainfed agriculture. Project designers clearly perceive that the real payoffs in such areas lie in irrigation development.

Projects that Focus on Rainfed Agriculture

In areas with limited opportunity for water harvesting, watershed projects typically devote more attention to developing rainfed agriculture. This is the situation in eastern Maharashtra and northern Andhra Pradesh, where the terrain is flatter and the climate less arid. Watershed projects in these areas promote on-site soil and water conservation measures that improve the resource base for rainfed agricultural production. This is intended to pave the way for adoption of crop varieties that are responsive to increased moisture. These projects often build water-

[2]Personal communication with Shri Anna Hazare 1996.

harvesting structures such as check dams and percolation tanks, but they cannot offer the spectacular increases in irrigation achieved in places like Ralegan Siddhi, because the terrain does not provide the same opportunities for harvesting water.

In southern Andhra Pradesh, the most obvious opportunities for water harvesting have long since been exploited in the form of traditional irrigation tanks. Some opportunities remain, but often they lie in the catchment of an existing tank, thus interfering with the traditional system. This helps explain why most projects in Andhra Pradesh focus more on rainfed agriculture than irrigation.

Projects Operating in the Study Regions

Government Projects that Focus Primarily on Water Harvesting

This discussion of the different watershed projects operating in the study area begins with the project sponsored by the government of Maharashtra, because the Maharashtra projects represent the roots of watershed development in India. The Jal Sandharan program is the result of several decades of experience with watersheds in the state.

Watershed projects in Maharashtra.[3] The beginnings of watershed development in Maharashtra date back to the 1942 Bombay Land Improvement Schemes Act. This initiative resembled modern watershed projects in its focus on soil and water conservation, improved rainfed farming methods, and controlled grazing. Watershed management gathered momentum in Maharashtra following the severe 1972 drought. Soon after, the government of Maharashtra launched the Employment Guarantee Scheme, which aimed to provide work to anyone who needed it while also creating permanent assets such as infrastructure. One important objective was to "drought-proof" the land by building water-harvesting structures that would provide drinking water and irrigation throughout the year.

In 1982, the government of Maharashtra initiated the Comprehensive Watershed Development Program (COWDEP). This program was intended to combine the budgetary resources of the Employment Guarantee Scheme and the technical provisions of the 1942 Bombay Land Improvement Schemes Act for a large-scale watershed development effort. But work undertaken by COWDEP was administered by several government departments, and coordination among them proved to be difficult.

Following COWDEP and other experiments in watershed development, the government of Maharashtra launched the Jal Sandharan Program in 1992. It represents an effort to take a more comprehensive approach to watershed development, with the key innovation being that the four government departments involved in the work were brought under one umbrella. The Jal Sandharan, which became a department in itself, would also handle the funds from the centrally sponsored Drought-Prone Areas Programme (DPAP), Jawahar Rojgar Yojana (JRY), and the National Watershed Development Project for Rainfed Areas (NWDPRA).

The Jal Sandharan program treats the village as the unit of planning, implementing the work in microwatersheds that lie within village boundaries. Emphasis is given to raising the water table to protect and enhance drinking water sources and provide protective irrigation for at least one crop. The program is implemented by a district-level committee representing all the government agencies involved in the project. The work in each selected village

[3]This discussion of state government programs in Maharashtra draws on Pangare and Gondhalekar 1998.

proceeds with the consent of the village *sarpanch* (elected leader) after a meeting of villagers is held to discuss the project.

The Jal Sandharan shows signs of lessons learned from several decades of state government experience in watershed development, but it also demonstrates clearly the difficulties of coordinating large-scale activities across government departments. In particular, coordination in the upper levels of bureaucracy does not always translate into coordination at the village level, where all the departments involved have separate budgets and targets (Pangare and Gondhalekar 1998).

Drought-Prone Areas Programme (DPAP). The DPAP is sponsored by the Ministry of Rural Development in the central government of India. The DPAP can be traced back to the Rural Works Programme initiated in 1971–72. It has evolved gradually over time, initially covering a wide range of labor-intensive activities such as soil and water conservation, revegetation and afforestation, and development of irrigation and infrastructure. Over time the program gradually focused more sharply on area development to prevent drought. By the late 1980s, the DPAP became exclusively a watershed development program focusing on soil conservation, water harvesting, pasture development, and afforestation. A small amount of funds were earmarked for associated activities such as livestock development, sericulture (raising silkworms), and horticulture.

As with other government-funded watershed programs, the DPAP was strictly a technical program in which local people played little or no role. Many nongovernment organizations (NGOs), meanwhile, had moved toward a more fundamentally participatory approach in which villagers shared in developing and implementing watershed plans. In 1994, India's Ministry of Rural Development adopted this approach on the basis of the well-known Hanumantha Rao Committee Report (India, Ministry of Rural Development 1994b). Under the guidelines subsequently drafted, plans were to be developed by the villagers, with an emphasis on the use of local technologies (India, Ministry of Rural Development 1994a). Funds would go directly to a village, with villagers working hand in hand with an independent project-implementing agency that could come from the government, nongovernment, or even the corporate sector. A strong effort was made to move away from the physical target orientation that characterizes most government programs. It took time to put this radical restructuring of the program into operation, and by 1997 almost no work had been undertaken in Maharashtra. Progress was better in Andhra Pradesh, but insufficient work had been done to warrant analysis of any project initiated after the guidelines were introduced As a result, the field work for this study covered villages under the DPAP using pre-1995 guidelines. In Maharashtra, the pre-1995 DPAP is synonymous with the COWDEP and Jal Sandharan, and in Andhra Pradesh the approach is very similar. In fact, before the implementation of the new DPAP guidelines the Jal Sandharan drew most of its budget from the DPAP.

Government Projects that Focus Primarily on Rainfed Agriculture

While the water harvesting and afforestation approach to watershed management was gathering momentum in Maharashtra and in the DPAP, alternate approaches were being introduced that focused more on developing rainfed agriculture through on-site soil and water conservation practices. These approaches were led in India by on-station research undertaken by institutes under the Indian Council of Agricultural Research (ICAR) such as the Central Research Institute for Dryland Agriculture (CRIDA), as well as the International Crops Research Institute for the Semi-Arid Tropics (ICRISAT), which is located in India.

ICAR model watersheds. In the mid-1980s, ICAR decided to implement the findings of its dryland agricultural research in 47 model watersheds around the country. Many of these pilot sites were treated as research watersheds, where the work undertaken was closely monitored and changes in land and water conditions were analyzed. Physical costs of the watershed works were relatively low, but supervision was intensive, with persistent efforts to introduce new varieties and other improved technologies and management practices.

CRIDA took up three such watersheds in Andhra Pradesh. One of them, Chevella, is included in this report. The other two, both of which were close to Hyderabad, could not be included because they have since been converted to housing developments. At least one model watershed was launched in Maharashtra, but it was in the eastern portion of the state not covered by the present study.

World Bank Pilot Project for Watershed Development in Rainfed Areas. This project was initiated in 1984 in Karnataka, Madhya Pradesh, Maharashtra, and Andhra Pradesh. Like the ICAR model watersheds, it sought to introduce improved rainfed agricultural technology. The project's guiding philosophy was that low-cost soil and water conservation measures, including improved agronomic practices like contour cultivation and vegetative rather than mechanical bunds, could make a strong contribution to rainfed agricultural development at a relatively low cost (World Bank 1988). While the work was undertaken on a watershed basis, the project also emphasized proper treatment *within* each plot, as the project's design team felt this was missing from the watershed approach pioneered in western Maharashtra. A major thrust of this project would be to promote contour-based cultivation, which would conserve soil and concentrate moisture at very little monetary cost. The improved soil moisture regime in turn would make improved seeds, fertilizers, and other inputs more productive.

The project also applied lessons learned from earlier projects regarding institutional approaches. Efforts were made to streamline state government operations to support the project; special offices were established at both the central and state levels in order to coordinate the administrative needs of the project.

Like other early government projects, the pilot project aimed for universal implementation of a single, centrally developed plan, with efforts made to convince local people of its merits. The project document stressed the need to adapt proven technologies to local conditions, but in practice there was little flexibility. Techniques not pre-approved under the project design were not supported. Recognizing that pasture development and afforestation work would have no lasting impact unless people supported it, the project mandated that no pasture development work would be undertaken without local people's consent. But that was the extent of efforts to gain local participation.

When the pilot project ended in 1991, a second phase of the World Bank project was introduced. This project was called the Integrated Watershed Development Project (IWDP), with separate components in the hills and plains. The plains portion of the project was undertaken in Rajasthan, Orissa, and Gujarat. The IWDP took essentially the same approach to developing rainfed crop production as the earlier pilot project. Its main difference was that it focused greater attention to developing and strengthening local organizations as the means of garnering people's participation and approval of collective action for protecting common pasture areas. The project's administrative approach was also restructured.

Project evaluations suggest mixed performance. In Rajasthan, the IWDP suffered from the same problems of poor participation and inflexible technology choice as the pilot project (ICRISAT 1996; Rajasthan Agricultural University 1999). In Orissa, on the other hand, the project is said to have performed well.[4] A second phase of the IWDP was commissioned for the Himalayan foothills in 1999; increased transfer of control to local community groups is an important stated goal (Vedeld 2000).

Since the present study is confined to Andhra Pradesh and western Maharashtra, the analysis addresses only the pilot project.

National Watershed Development Project for Rainfed Areas (NWDPRA). This program was initiated as the Ministry of Agriculture's counterpart to the World Bank-funded pilot project and IWDP. Similar in approach to the World Bank projects, the NWDPRA promoted the same low-cost vegetative bunding techniques and contour-based cultivation (India, Ministry of Agriculture 1991). Vegetative and other low-cost measures were also used on nonarable lands. The NWDPRA is centrally funded and operates through state-level agriculture or watershed development departments. In 2000, the NWDPRA adopted the Ministry of Rural Development's more participatory guidelines, and in the coming years the two projects are to be implemented with common guidelines (India, Ministry of Agriculture 2000). This analysis covers villages developed under the NWDPRA's earlier approach.

Another similarity to the World Bank projects is that the NWDPRA works on a watershed basis, where watersheds do not necessarily correspond to village boundaries. The NWDPRA watersheds are smaller, ranging from 500 to 5,000 hectares, or about 5 to 20 percent of the area of the World Bank watersheds. As a result, the NWDPRA watersheds typically cover one village entirely or nearly so, plus parts of one or two neighboring villages. This approach is considered to make the most sense from a land and water management perspective, but it raises administrative complications because project staff has to deal with multiple village administrations in one relatively small area. Also, organizing local institutional arrangements for managing nonarable common lands is complicated when working across village boundaries. The World Bank pilot project and IWDP shared this problem and had difficulty making it work (ICRISAT 1996).

The NWDPRA project guidelines mentioned the issues of people's participation and institution building, but they presented no clear strategy and only a small budget for addressing them (India, Ministry of Agriculture 1991). It appeared that the project's intentions were in line with modern views about the benefits of participation but that the mechanisms for ensuring them were not fully developed.

In western Maharashtra, the NWDPRA was implemented by the same agency responsible for the engineering-based approaches of COWDEP, Jal Sandharan, and DPAP. The primary focus remained on treating drainage lines and catchment areas to promote infiltration of water. But the technologies used in the NWDPRA project were much less expensive. For example, drainage line structures under the NWDPRA contained no cement and were limited to a maximum cost of Rs 25,000 per structure; under the other projects, individual water harvesting structures might cost seven or eight times as much. As a result, water harvesting was not the NWDPRA's strength.

In Andhra Pradesh, the NWDPRA was operated by the Department of Agriculture and more clearly matched the approach envisioned in the project guidelines.

[4]Personal communication with World Bank officials involved with the project.

NGOs: A Focus on Social Organization

NGO programs are by no means uniform, but they all place strong emphasis on social organization. Their guiding principle is that without proper social organization, efforts to introduce watershed technology will be fruitless.

The two features that most distinguish NGO watershed programs from government programs are their scale of operations and their staffing structure. While government programs have huge budgets and work in hundreds of villages, most NGOs work in only a handful of villages. They devote more staff time per village, and they often work on a variety of activities in addition to watershed management. While government employees involved in watershed management are almost exclusively trained in agricultural sciences and engineering, NGO staff members include many more nontechnical staff trained in community organization. They believe that social organization contributes as much to successful watershed development as technical input. Some NGOs collaborate with government agencies that provide technical expertise, but others do not. Some NGOs also aim to help villagers gain access to a variety of government services such as health care and education.

It is important to note that NGOs vary a great deal. Some are large and well established, with access to substantial funding, whereas others are small, less experienced, and underfunded.

NGOs in Maharashtra.[5] Watershed management in Maharashtra has roots in the nongovernment sector that go back nearly as far as the government programs. In the early 1980s two villages became well known for their watershed management programs: Ralegan Siddhi in Ahmednagar district and Adgaon in Aurangabad district. Many current government and NGO initiatives draw inspiration from these successes.

In the 1970s, Ralegan Siddhi was a poorly developed village almost devoid of trees and grass and a haven for liquor dens. Anna Hazare emerged as a local leader. He brought about various social changes in the village, particularly family planning, a ban on alcohol, protection of nonarable lands against open grazing and felling of trees, and *shramdan*, or voluntary labor for community welfare. Around the same time he also learned about the benefits of soil conservation and water harvesting. The social changes brought order and a sense of community to the village, while soil and water conservation work (implemented by COWDEP) and protection of the common lands helped restore the natural resource base. This was the beginning of people's participation in watershed development.

Among the many NGOs working in watershed development in Maharashtra, one of the best established is Social Centre, which was founded in Ahmednagar in 1969 by Jesuit priests. During the period 1969–88, it was engaged in various activities such as small loans, community lift irrigation schemes, and community health programs. In 1988 it shifted its focus toward motivating and organizing entire villages to undertake ecological regeneration of their own watersheds. Social Centre played a key role in launching and designing the statewide Indo-German Watershed Development Programme, discussed later in this chapter.

NGOs in Andhra Pradesh. NGOs in rural Andhra Pradesh have traditionally focused on the problems of scheduled castes, backward classes, and scheduled tribes.[6] Caste structure is more dichotomized in Andhra Pradesh than in Maharashtra, with more villages in which one or two large landowning families control large tracts of land while many families are landless. As a

[5]This discussion of NGO projects in Maharashtra draws on Pangare and Gondhalekar 1998.

[6]These are the official terms for lower caste and tribal communities.

result, NGOs typically focused on activities that were not land-based, such as developing and strengthening local credit institutions. With the rise of watershed development as a focal point for rural development, some NGOs gradually adopted it into their project portfolio.

In recent decades Andhra Pradesh has had successive waves of large-scale privatization of common lands, wherein landless and near landless people were given legal but nontransferable title to common lands (Pender and Kerr 1999). Many NGOs expanded their credit and other income-generation activities to support agriculture on the privatized land, much of which is of low quality. When they expanded into a watershed approach, they also began to work with farmers with higher quality land. But as a result of their primary orientation toward helping poor, landless people, these watershed agencies tend to be committed to making landowners pay for work done on their own property. For example, while most projects in Maharashtra and the centrally funded government programs typically ask for no more than a 10 percent contribution from farmers for work done on their private lands, some NGOs in Andhra Pradesh require a more substantial contribution. Some of the implications of this policy are discussed in the section on land improvement investments.

MYRADA, a Karnataka-based NGO operating in Andhra Pradesh, focuses on building consensus among different interest groups in a watershed to a much greater extent than Maharashtran NGOs. Because the costs and benefits of watershed development are often spread unevenly, project implementation can be difficult, especially where socioeconomic diversity is high. MYRADA addresses this problem by helping communities develop mechanisms to compensate those who lose, so that they will cooperate for the greater good (Mascarenhas et al. 1991; Fernandez 1993; Fernandez 1994).

In all of its rural development projects, MYRADA organizes people into small, homogeneous groups working toward one common purpose. In the context of watersheds, they start with "miniwatersheds" of no more than a few hundred hectares and a hundred farmers. Within the miniwatershed, MYRADA helps form small subgroups of farmers based on homogeneity of location, socioeconomic conditions, or interests. These groups all belong to a larger microwatershed group. This preserves the participatory and socially functional character of the smaller, homogeneous subgroups, while also retaining advantages of scale in planning watershed works and interacting with government agencies, banks, and input suppliers. The larger group provides a vehicle for airing complaints and settling disputes among people from different subgroups.

NGO/Government Collaborative Programs

The most interesting aspect of watershed development in Maharashtra in recent years is the rise of collaborative programs between government and nongovernment agencies. The two main examples are the Adarsh Gaon Yojana (Ideal Village Scheme, or AGY), and the Indo-German Watershed Development Programme (IGWDP), funded by the German government. The new guidelines of the DPAP also offer opportunities for government agencies and NGOs to collaborate.

Adarsh Gaon Yojana(AGY).[7] The AGY is a major initiative that seeks to replicate the Ralegan Siddhi model in 300 villages by combining the technical staff of the Jal Sandharan program with the social orientation of NGOs.

[7]This discussion of the AGY draws on Pangare and Ghondhalekar (1998).

The key elements of the AGY are collaboration between an NGO and the government and strict guidelines for social organization. Villages participating in the AGY must undertake to follow the five social principles of Ralegan Siddhi: family planning, a ban on alcohol, a ban on open grazing, a ban on cutting trees, and voluntary community work *(shramdan)*. It is thought that adherence to these five principles can point a village toward self-sufficiency by helping villagers meet their needs for water, food, fuel, and fodder within their own villages. The philosophy also promotes a set of values that encourages self-discipline and a willingness to overcome social barriers and political factionalism to work for the common good.[8]

Shramdan is intended to foster a spirit of self-sufficiency and interdependence. In theory, when villagers observe the benefits of the physical works they have helped create for watershed development, they have a sense of satisfaction and achievement. They also feel responsible for maintaining the structures in which they have invested their own labor. *Shramdan* is also seen as a good way of getting people together to work for the welfare of the entire community.

NGOs play an important role in the AGY. People in each village select a local NGO to help them implement the different development activities and adhere to the social principles. The NGO also maintains records and accounts and monitors the project activities. In addition, the NGO coordinates with the government departments at the state level to access funding and technical guidance. The Jal Sandharan Department, meanwhile, implements the technical work.

Funds under the project are to be used for two main types of activities: watershed development (the core activity) and other development activities (noncore activities). The latter are carried out by the appropriate government agency, in consultation with the people of the village. Government departments are supposed to give AGY villages preference in providing services, taking steps to reduce corruption and peripheral expenses.

Indo-German Watershed Development Programme (IGWDP). The IGWDP, another example of collaboration between government and nongovernment organizations, seeks to replicate the success of small NGO programs at a larger scale (Farrington and Lobo 1997; WOTR 2000; NABARD 1995). Initiated in 1993, the IGWDP develops microwatersheds in a comprehensive manner through the initiative taken by village groups. Its guiding philosophy is the need for collaboration among village-level organizations, NGOs skilled in social organization, and government organizations skilled in technical work. Further, it accepts that although indigenous knowledge and practices are important, they need to be augmented by modern techniques and management practices. The IGWDP has developed elaborate procedures to cut through bureaucratic turf wars and red tape, ensuring that funds move quickly (Farrington and Lobo 1997). As of June 2002, the IGWDP had undertaken development programs in 146 villages, covering about 137,000 hectares and involving 78 NGOs (WOTR 2002). Plans are being considered to spread this program to other states.

Investment in physical capital under the IGWDP begins only after evidence of social organization suggests that people will work

[8]It is important to note that *shramdan* has a long history in Maharashtra and is considered culturally appropriate. In other areas, other means of promoting cooperation and social discipline may be preferred. In southern Rajasthan, for example, Seva Mandir insists that villagers reverse all illegal encroachment on common lands before they will undertake work there (Seva Mandir 1999; Ahluwalia 1997).

together to maintain the investments on both private and community land. As with the AGY, the emphasis on developing the village's social capital is as strong as that on developing its natural and physical capital, and the villagers must submit to similarly strict social conditions.

The work begins with 12 to 18 months of social organization work. This is almost 12 to 18 months longer than the social organization phase of a typical government watershed program at the time of the survey, but it is shorter than that of many NGOs, which conduct work on several other areas of village development before venturing into watershed development. One important early project activity under the IGWDP is to plant trees and grasses in the catchment area. This is done prior to building water harvesting structures in order to force the inhabitants of the village to show that they can enforce social fencing—self-policing without the use of fences—to protect natural vegetation. Only after people demonstrate the ability to discipline themselves does the project invest larger amounts of funds in new watershed structures.

The NGO helps organize and develop a Village Watershed Committee, which is essentially a village-based NGO. The idea is that the committee will eventually outgrow the need for support from the original NGO.

1994 guidelines for the Drought-Prone Area Programme (DPAP). The Ministry of Rural Development's 1994 guidelines for the DPAP and some of its other programs represented a radical change from the earlier technocratic, top-down approach. The new guidelines, which the Ministry of Agriculture also adopted in 2000 (India, Ministry of Agriculture 2000), drew heavily on the experiences of various high-quality NGOs. Not surprisingly, in practice the guidelines resemble the approaches taken by the NGO and NGO/government collaborative projects. In an effort to make the work more demand-driven and to create pressures on government agencies to improve their performance, the DPAP funds watershed development through either government or nongovernment agencies. As with the IGWDP and AGY, the new guidelines aim to create conditions whereby the technical strengths of government organizations and the community-building strengths of NGOs can be combined.

Another innovation was to move away from the typical target orientation of most government projects. For example, in recognition that watershed development may not be villagers' highest priority or that steps may be required to facilitate collective action, the new guidelines provide funds for entry-point activities that may have nothing to do with watershed development per se but that help address villagers' felt needs and generate enthusiasm for the watershed project (Turton et al. 1998; India, Ministry of Agriculture 2000.) In the early years the entry-point activity had to be completed quickly, and it often was viewed as just another target to be met. Subsequently, the guidelines were revised so that up to a year could be devoted to the entry-point activity and community organization phase. This is similar to the IGWDP's approach of devoting up to 18 months to community organization.

As already explained, project performance under the new guidelines could not be analyzed in this study because they were still in the early stages of implementation at the time the fieldwork was conducted. However, analysis of the performance of NGO and NGO/government collaborative projects is likely to shed light on the prospects for the new guidelines because of the similarity in their approaches.

Investment Costs per Hectare Under Each Project

Information about the cost per hectare under different projects is helpful in assessing their cost effectiveness. It also helps in interpreting the findings of quantitative analysis presented later in this report; one

Table 2.1 Estimated cost per hectare of watershed development under different programs

	Approximate cost per hectare (1998 Rs.)	
Prject Category	Excluding COWDEP	Including COWDEP
NWDPRA	2,500–3,500	4,000–6,000
Jal Sandharan and pre-1994 DPAP	2,500–4,000	4,000–6,500
Post-1994 DPAP	4,000	n.a.
NGO	4,000–6,000	5,500–8,500
AGY or IGWDP	3,500–5,500	5,000–8,000
World Bank	5,500–6,500	n.a.

Source: Authors' 1997 survey data

Notes: See Chapter 2 or the glossary at the end of the report for full names and descriptions of the projects. N.a. indicates not available.

should not be surprised if a project that spends twice as much as another has greater measurable impact.

Unfortunately, measuring project costs is difficult. Some projects, including the NWDPRA and the World Bank projects, have expenditure guidelines that can be taken as a broad indication of the level of investment per hectare. For other projects, estimates must be calculated based on the total expenditure and the total area covered, but records are difficult to obtain. For example, officials of the Jal Sandharan project say that their budgets are constructed on the basis of structures to be built and vegetation to be planted, not the area of the watershed. Accordingly, calculations of cost per hectare are only approximate.

NGOs tend to keep poor records of costs per hectare, and calculating them is extremely difficult because NGOs tend to undertake a variety of activities in addition to watershed development. So, even when costs can be calculated, what they cover is not always clear. One certainty is that NGOs have higher staff costs than government projects, since they devote much more time to social organization for which expenditures are not directly tied to treated area.

In Maharashtra, where nearly all projects operate in ex-COWDEP villages, calculating costs requires taking the sum of expenditures under both the old and new programs. Records from the old projects are poor, so the cost figures are only approximate.

Rough estimates of project costs per hectare by project category are presented in Table 2.1. For the NWDPRA and the World Bank Pilot Project the upper range is the cost listed in the project guidelines; for the post-1994 DPAP, it is exactly the amount listed in the guidelines; while for the Jal Sandharan and COWDEP, it is based on the total number of structures built divided by the area covered. AGY and IGWDP costs are calculated similarly but there are higher staff costs. The NGO figures are based on estimates provided by the officials interviewed.

CHAPTER 3

Data

Evaluating watershed projects requires baseline and monitoring data for comparison of conditions in the watershed before and after the project, but unfortunately no such information was available for this study of projects in Maharashtra and Andhra Pradesh. As a result, the quantitative analysis is based on secondary data available for the period before the project (1987[9]) and the time of the survey (1997), primary data of 1997 conditions based on interviews and visual assessments, and primary data of past conditions based on recall by local inhabitants. Inevitably there are weaknesses in the data that limit the study's analytical power.

A major component of the research was the development and collection of data on various indicators of performance in natural resource conservation, agricultural productivity, and equitable distribution of project benefits. These data were collected through direct observation, group discussions, and published records. Quantitative data were also collected on the background characteristics of the projects, villages, households and plots covered under the study. Some of the village-level information came from public sources, but most of it was collected from group and individual interviews in each village. In addition, qualitative data were collected regarding the natural resources people use to earn their livelihoods, the social institutions that govern access to those resources, and any changes in access resulting either from changes in the quantity of a resource or in the social institutions. This information was collected in open-ended discussions with members of specific interest groups in each village, such as farmers with irrigated land, farmers with rainfed land, landless people, herders, and women.

The village, rather than the watershed, was selected for analysis of community level indicators of natural resource management and economic performance. This is because most projects in the sample worked at the village or subvillage level, people are organized around villages, and secondary data are recorded at the level of the village. In some cases, particularly in Andhra Pradesh, villages are disaggregated into hamlets, in which case primary data were collected at the hamlet level.

---

[9]Work in the World Bank and ICAR project villages began in 1986; in villages under these projects the baseline period was the year before the project began.

Project Categories Covered in the Analysis

The research covers all types of projects that had been operating in Andhra Pradesh and Maharashtra for several years at the time the data were collected. They include the following categories:

- *National Watershed Development Project for Rainfed Areas (NWDPRA) projects*
- *Indian Council of Agricultural Research (ICAR) Model Watershed Projects*
- *World Bank-assisted Pilot Project for Watershed Development in Rainfed Areas*
- *Jal Sandharan, with funding from the Ministry of Rural Development under the pre-1994 guidelines*
- NGO projects
- NGO/government collaborations
- Nonproject: villages with no watershed project.

All of these project categories were discussed in detail in Chapter 2. The NGO/government collaborative projects were found only in Maharashtra, whereas the World Bank and ICAR projects were found in both states but not in western Maharashtra where the field work was conducted. The more recent World Bank-assisted IWDPs did not operate in either Maharashtra or Andhra Pradesh, so they are not covered in the quantitative analysis. The post-1994 DPAP is not covered because insufficient progress had been made at the time the field work was conducted. However, findings related to the performance of NGO and NGO/government collaborative projects can offer insights into the post-1994 DPAP because they take a similar approach.

Performance Indicators

There is no single indicator of successful watershed development, so the most feasible approach is to compare the performance of a variety of indicators, which also reflect the diversity of project objectives. These include, among other things, raising rainfed agricultural productivity, recharging groundwater for drinking and irrigation, raising productivity of nonarable lands, reducing soil erosion, skewing benefits toward poorer members of society, creating employment (directly and indirectly), promoting collective action, and building or strengthening social institutions. All the projects surveyed shared most of these objectives but, as described in Chapter 2, they differed in their relative emphasis.

The indicators vary in their level of rigor and reliability, which is inevitable given the lack of baseline or monitoring data in the study villages. Table 3.1 presents an overview of performance criteria, ideal indicators, and the indicators actually used in the current study, and Table 3.2 shows the level at which they operate.

Determinants of Project Performance

Village level. Data collected at the village level are based on a survey covering background information such as access to markets, land use patterns, natural resource management practices, and description of social institutions operating in the village. Most background information is available for both 1987 and 1997. Most of this information was obtained from a village-level survey conducted jointly by IFPRI and NCAP, and additional background variables were obtained from the 1991 census. Performance indicators at the village level include some variables from the village survey, but also visual observations of natural resource conditions obtained by walking through the village's farmland and uncultivated land. These transects covered a cross-section of broadly representative land types and uses.

Plot level. A plot-level survey was conducted to collect data on agricultural productivity and adoption of improved tech-

Table 3.1 Performance indicators used to compare project performance

Performance criteria	Ideal indicators[a]	Proxy indicators used in this study
Soil erosion	Measurement of erosion and associated yield loss	Visual assessment of rill and gully erosion (current only)
Measures taken to arrest erosion	Inventory, adoption, and effectiveness of soil and water conservation (SWC) practices	Visual assessment of SWC investments and apparent effectiveness (current only)Adoption of conservation-oriented agronomic practices Expenditure on SWC investments
Groundwater recharge	Measurement of groundwater levels, controlling for aquifer characteristics, climate variation, and pumping volume	Approximate change in number of wellsApproximate number of wells recharged or defunct Change in irrigated areaChange in number of seasons irrigated for a sample of plots Change in village-level drinking water adequacy
Soil moisture retention	Time series, intrayear, and interyear variations in soil moisture, controlling for climate variation	Change in cropping patterns Change in cropping intensity on rainfed plots Relative change in yields (higher, same, or lower)
Agricultural profits	Net returns at the plot level	Net returns at the plot level, current year only
Productivity of nonarable lands	Change in production from revenue and forest lands (actual quantities)	Relative change in production from revenue and forest lands (more, same, or less than pre-project) Extent of erosion and SWC on nonarable lands
Household welfare	Change in household income and wealth Nutritional status	Perceived effects of the project on the household Perceived change in living standard (better, same, worse) Change in housing quality Change in percentage of families migrating Perceived changes in real wage and availability of casual employment opportunities (higher, same, lower)

Source: Devised by authors.
Note: [a] Ideally, all indicators would be collected both before and after the project.

nologies and practices. This survey provides information about changes resulting from the watershed projects and other determining factors. The sample includes both irrigated and rainfed plots and plots covered and not covered by watershed projects. Village-level information related to each plot is available from the village survey. Some household-level information for each plot was also collected as a part of the plot survey.

Data from the household survey and interest group interviews are mainly related to indicators of performance rather than determining factors, so they are not discussed here.

Sampling

Sampling villages for data collection was a major undertaking in itself. The situation in Pune and Ahmednagar districts of western Maharashtra provides a good example of the difficulties. Despite widespread publicity about the success of the watershed approach to agricultural development, hard data were quite limited. With the exception of a few widely known success stories, project villages were not easy to locate for two reasons. First, the 1991 Census lists more than 3,000 villages in the two districts: the successful villages accounted for no more than a handful. Second, a complete list of villages where projects have

Table 3.2 Performance criteria and the level of analysis at which they operate

Type of performance	Level of analysis			
	Village	Interest group	Household	Plot
Social organization	Institutions to protect common lands Use of voluntary community labor	Village organizations	Membership in village organization	...
Natural resource conservation	Irrigated area Drinking water supply Soil erosion and conservation Products from common lands Habitat for wild animals and migratory birds	Soil conservation	Drinking water supply	Irrigated area Population of improved dairy cattle
Agricultural productivity	Soil erosion and conservation	Irrigation Rainfed agriculture	...	Irrigation Cropping intensity Adoption of new varieties Crop yield Returns to cultivation
Equity and poverty alleviation	...	Access to resources Access to employment	Assets (wealth) Access to employment Standard of living Condition of housing	...

Source: Devised by authors.

operated did not exist. The most active watershed agency in the area, the Maharashtra Department of Soil and Water Conservation, kept good records of the villages where work was currently under way, but lists of villages where work had been completed were archived and difficult to access. Some government programs, like the NWDPRA, maintained lists of project locations only at the *taluka* level[10]. NGOs maintained their own lists, which could be obtained by visiting the head office. As a result, simply identifying project villages required a great deal of legwork. The resulting list of project villages was then checked against the complete list of all villages from the national census so that non-project villages could be selected as a control against which to compare project performance. All the sampled villages were visited to confirm their project status.

The Maharashtra study villages were all located in Pune and Ahmednagar districts in the western part of the state, where there was a relatively high concentration of watershed project sites. The eastern side of this study area is drought-prone, while the area closer to the Western Ghat mountain range has higher rainfall. In Andhra Pradesh the projects were less concentrated, so the sample villages covered four districts: Anantapur in the far south of the state and Medak, Mahbubnagar, and Ranga Reddy, all of which are in the north, near Hyderabad.

[10]Talukas are small administrative units that make up districts.

Based on available knowledge about project status of villages, the sample was selected at random, stratified by the listed project categories, and, in Maharashtra, by geographic location. With five project categories and two geographical zones, there are 10 strata in Maharashtra. A small amount of resampling was done to replace villages incorrectly classified as "nonproject" after visits to the villages revealed that watershed projects had operated there in the 1980s. In Andhra Pradesh, where only 16 villages were sampled, geographic stratification was omitted to ensure that each stratum had at least two observations. Thus Andhra Pradesh has only 5 strata.

Data were collected in 70 villages in Maharashtra and 16 villages in Andhra Pradesh (Table 3.3). A full set of quantitative and qualitative data at the village, household, and plot level were collected in a random, stratified subsample of 13 of the Maharashtra villages and all 16 Andhra Pradesh villages, for a total of 29. In the remaining 57 Maharashtra villages only village level data were collected. The village-level analysis is confined to the 70 Maharashtra villages, while the plot-level analysis covers the 29 villages from both states where more detailed data were collected. The qualitative data cover primarily these same 29 villages.

Teams of five to seven village investigators spent four days and nights in each of the 29 villages where they collected the full set of quantitative and qualitative data. In the remaining 57 villages in Maharashtra where only village-level data were collected, teams of three to four investigators spent two to three days.

Characteristics of the Sampled Villages and Plots

Village Characteristics

Village-level analysis is presented only for Maharashtra. Conditions in the two states are sufficiently different to make it useful to analyze them separately, and the Maharashtra sample of 70 villages facilitates the analysis.

Villages are characterized by a variety of factors that might affect performance in agricultural productivity, natural resource management, and living standards. Table 3.4 defines the variables representing these characteristics.

Table 3.5 shows the number of sampled villages in each project category and also the total number of villages in the study area. The number of sampled villages under the NWDPRA, NGO, and AGY and IGWDP categories is very small, reflecting the fact that these projects were not very widespread at the time of the study. The sample includes nearly the entire population for the projects,[11] whereas the Jal

Table 3.3 Number and location of the study villages

Level of analysis	Maharashtra	Andhra Pradesh	Total
Total	70	16	86
Village level (quantitative)	57	0	57
Village, plot, household (quantitative), and interest group (qualitative)	13	16	29

Source: Authors' 1997 survey data.

[11]Subsequent expansion of the IGWDP and DPAP raised the number of villages in these categories. Under its new guidelines, the DPAP works mainly through NGOs, so the number of NGO-led projects in the study area is slated to rise quickly.

Table 3.4 Definitions of variables for analysis at the village level

Description of variable	Years for which data are available	Source of data
Indicators of performance		
Change in % of area irrigated between 1987 and 1997	1987-97	Government statistics
Overall rill erosion status by land use type	1997	Visual observations from the transect survey
Soil conservation investment status by land use type	1997	Visual observations from the transect survey
Overall condition of drainage line	1997	Visual observations from the transect survey
Extent and conditions of bunds lining the drainage line	1997	Visual observations from the transect survey
Extent of breaches in the sides of the drainage line	1997	Visual observations from the transect survey
Condition of junctions between drainage lines in the main drainage line	1987-97	Visual observations from the transect survey
Change in % of families migrating for at least one month	1987-97	Primary data
Change in availability of fodder and fuel from the common revenue land (Rs/hectare)	1987-97	Primary data
Determinants of performance of the project and/or selection of a village for inclusion in a project		
Location: state, district	...	Government statistics
Altitude range of the village from highest to lowest point (meters)	...	Government statistics
Position in the macrowatershed (lower, middle, or upper reaches)	...	Government statistics
% of village area under Forest Department	1987-97	Primary data
Distance to taluka headquarters (kilometers)	1987-97	Primary data
Distance to nearest bus stop (kilometers)	1987-97	Primary data
Visits by an extension agent (number per year)	1987-97	Primary data
Number of communities in the village	1997	Primary data
% of villagers who are from SC, ST, and BC communities[a]	1990	National census
% of villagers in different occupations	1990	National census
Distance to bank (kilometers)	1987 and 1997	Primary data
Restrictions against grazing on common revenue land (yes/no)	1987 and 1997	Primary data
Illicit grazing is punished (yes/no)	1987 and 1997	Primary data
Shramdan (voluntary communal labor) is practiced in the village (yes/no)	1987 and 1997	Primary data
History of development projects that have operated in the village	1987 and 1997	Primary data
Strong leader who promotes social and economic development (yes/no)	1997	Primary data
Watershed project operating in the village	1987 and 1997	Agency records and primary data
% of the village actually covered by the project	...	Agency records
Money spent by the project (Rs/hectare)	...	Agency records
Year project began and ended	...	Agency records
Population density (inhabitants per square km)	1990	National census
Distance to nearest industrial unit	1987 and 1997	Primary data
Distance to regulated market (kilometer)	1987 and 1997	Primary data
Informal credit groups (yes/no)	1987 and 1997	Primary data
Number of functional water-harvesting structures	1997	Primary data
Social restrictions on water use for irrigation (yes/no)	1987 and 1997	Primary data
Nominal daily wage (Rs)	1987 and 1997	Primary data
Average annual rainfall at the nearest taluka headquarters (millimeters)	...	Government records

Note : Primary data from 1987 are based on respondents' recall.
[a]SC is scheduled castes, ST is scheduled tribes, and BC is backward classes.

Sandharan and nonproject villages in the sample represent only a small fraction of their total populations.

Project Characteristics

Projects vary greatly in characteristics such as the percentage of village area they cover, the number of years they operate, the amount of funds they spend, and the training of their staff. Table 3.6 presents the approximate mean values of such information for the sampled villages in Maharashtra. The percentage of each village covered and the total expenditure per hectare in each village[12] are slightly less under the NWDPRA than under other projects, but these

Table 3.5 Number of villages in the Maharashtra village-level survey, by project category

Prject Category	Number of villages sampled	Total number of villages in the study area
NWDPRA	10	11
DPAP and Jal Sandharan	17	201
NGO	12	13
AGY or IGWDP	14	27
No project	17	361
Total	70	613

Source: Authors' 1997 survey data.
Note: See Chapter 2 or the glossary at the end of the report for full names and descriptions of the projects.

Table 3.6 Project operations in villages under each project category, Maharashtra

Item	NWDPRA	DPAP and Jal Sandharan	NGO	AGV or IGDWP	No project
Mean area of the village (hectares)	2,102.0	1,422.0	1,209.0	1,144.0	905.0
Number of villages in which COWDEP previously worked (out of total sample)	9/10	16/17	11/12	14/14	0/17
Mean % of area covered by the old project (COWDEP)	38.0	49.0	43.0	46.0	...
Mean % area covered by new projects	36.0	39.0	42.0	38.0	...
Mean % area covered by both old and new projects	74.0	88.0	85.0	84.0	...
Mean % of staff members with training in social organization[a]	0	0	42.0	45.0	...
Mean number of years since the most recent project ended	0	0	0.3	0	...
Mean number of years of work under COWDEP	7.2	8.4	7.0	8.7	...
Mean number of years under the new project	5.9	4.9	6.0	5.1	...
Mean number of years under both projects	13.1	13.3	13.0	13.8	...
Mean cost/hectare actually treated under new project (Rs)	4,500.0	4,783.0	4,989.0	4,963.0	...
Mean expenditure/hectare for entire village under COWDEP (Rs)	1,880.0	2,355.0	2,148.0	2,310.0	...
Mean expenditure/hectare for entire village under new project (Rs)	1,622.0	2,006.0	2,207.0	1,874.0	...
Mean expenditure/hectare for entire village under both projects (Rs)	3,501.0	4,361.0	4,355.0	4,185.0	...

Source: Authors' 1997 survey data.
Notes: COWDEP is the Comprehensive Watershed Development Program initiated by the government of Maharashtra. Most figures are approximate, based on calculations from official records.
[a]Analysis of variance (ANOVA) test shows that group means differ significantly across project category (excluding nonproject) only for the percentage of staff members with training in social skills ($F = 45$, 3 df, $p < .001$). No other variables listed in this table show significant difference across project types.

[12]Total expenditure per hectare is calculated as the cost per hectare treated multiplied by the percentage of the village covered by the project.

Table 3.7 Number of villages covered in the plot survey, by state and project category

	Number of villages	
Prject Category	**Maharashtra**	**Andhra Pradesh**
World Bank or ICAR	0	4
NWDPRA	2	3
Jal Sandharan and DPAP	2	3
NGO	3	3
AGY or IGWDP	3	0
No project	3	3
Total	13	16

Source: Authors' 1997 survey data.
Note: See Chapter 2 or the glossary at the end of the report for full names and descriptions of the projects.

Table 3.8 Number of plots in the plot survey, by irrigation and land capability classification

	Land capability classification			
Irrigation status	**II**	**III**	**IV**	**Total**
Rainfed	22	99	125	246
Irrigated	37	59	9	105
Total	59	158	134	351

Source: Authors' 1997 survey data.
Notes: The land capability class is a system of categorizing land on the basis of its agricultural production potential, ranging from I (most fertile) to IX (largely rock). The cultivated land in the survey area ranged from class II to class IV.

Table 3.9 Sample for the plot survey, categorized by landholding size

Prject Category	Hectares operated	Number of plots
Small	0-2	118
Medium	2-4	114
Large	>4	119
Total	...	351

Source: Authors' 1997 survey data.

differences are not significant. The reader must keep in mind that the NWPDRA projects cover multiple villages, while the other projects work in only one village at a time.

Villages are also differentiated by the amount of time in which they have been involved in watershed projects. This includes the current or most recent watershed project and earlier work (if any) in the 1980s done by the Maharashtra Department of Soil and Water Conservation under COWDEP. By either measure, the average duration of projects is not significantly different across watershed project categories (Table 3.6). With the exception of the nonproject villages, all but two villages in the sample also had work done under COWDEP in the 1980s. The amount of work actually performed by COWDEP varied by village, but the figures in Table 3.6 are still striking: in one form or another, watershed development has taken place in these villages for a long time.

Finally, perhaps the most noticeable point in Table 3.6 is the difference across projects in the percentage of staff members trained in social organization skills. In the two government programs (NWDPRA and Jal Sandharan), no staff members had any training in social organization, while nearly half of them did in those with an NGO component. This reflects their differences in orientation toward social organization versus transfer of technology.

Plot Characteristics

Approximately 12 plots were sampled in each of 13 villages in Maharashtra and 16 in Andhra Pradesh. The villages were stratified by watershed project category; the resulting sample of villages selected is shown in Table 3.7. Within each village, plots and households were selected at random, stratified by land capability classification and irrigation status. Every third plot was selected from the transect line that crossed representative areas of the village, until four plots were selected in each of three land capability classes, with three rainfed plots

Table 3.10 Plot, household, village, and project characteristics that potentially determine performance at the plot level

Description of variable	Years data are available	Source of data
Biophysical and management characteristics of the plot		
Plot area (hectares)	1987 and 1997	Primary data
Average annual rainfall measured at the nearest *taluka* headquarters (millimeters)	1987 and 1997	Government data
Land capability classification: sample includes class II, III, and IV plots	1987 and 1997	Investigator's visual assessment
Irrigation status: 1 if it is irrigated at least one season, 0 if it is rainfed	1987 and 1997	Primary data
Slope: 1 if slope is 0-2%, 2 if slope is 2-4%, 3 if slope is 4-8%	1997	Primary data
Characteristics of the farm household		
Total hectares of land owned by the household	1987 and 1997	Primary data
Total number of adult workers in the household (men, women, and, long-term hired workers)	1997	Primary data
Highest number of years schooling of any male household member	1987 and 1997	Primary data
Highest number of years schooling of any female household member	1987 and 1997	Primary data
% of income from off-farm sources	1997	Primary data
Change in % off-farm income between 1987 and 1997		Primary data
Tenure status: 1 if the farmer owns the plot, 0 if the operator is a tenant or sharecropper	1997	Primary data
Land title status: 1 if the farmer has a transferable title, 0 if not	1997	Primary data
Plot rank (relative to farmer's other plots): 1 if it is the farmer's only plot or a better than average quality plot in his holding, 2 if it is an average quality plot, 3 if it is below average	1997	Primary data
Farmer interacted with project staff: 1 if yes, 0 if no	Anytime between 1987 and 1997	Primary data
Project staff made technical recommendations to farmer: 1 if yes, 0 if no	Anytime between 1987 and 1997	Primary data
Farmer adopted technologies or practices recommended by project staff: 1 if yes, 0 if no	Anytime between 1987 and 1997	Primary data
Village level		
Village, *taluka, district, state*		
Altitude range between highest and lowest point in the village (meters)	...	Government data
Distance to *taluka* headquarters (kilometers)	1987 and 1997	Government data
Distance to district headquarters (kilometers)	1987 and 1997	Government data
Distance to a large city (Pune for Maharashtra, Hyderabad, or Bangalore for Andhra Pradesh), kilometers	1987 and 1997	Government data
Type of road connecting the village (highway, paved road, good unpaved road, bad unpaved road, bullock cart path)	1987 and 1997	Primary data
Distance in kilometers to nearest bus stop	1987 and 1997	Primary data
Number of visits to the village by an extension agent per month	1987 and 1997	Primary data
Distance to the nearest bank (kilometers)	1987 and 1997	Primary data
Strong leader in the village promotes social and economic development (1 if yes, 0 if no)	1997 only	Primary data
Population density (inhabitants per square kilometer)	1990	1991 census
Distance to nearest industrial unit (kilometers)	1987 and 1997	Primary data
Distance to nearest regulated market (kilometers)	1987 and 1997	Primary data
Nominal daily wage (Rs)	1987 and 1997	Primary data
% of houses in the village with an electrical connection	1987 and 1997	Primary data
Project level		
Type of watershed project operating in the village, if any	1987-1997	Project records
Combined number of years under the most recent watershed project and another previous project	1997 and prior	Project records
Approximate percentage of the village's area covered under the project	1997 and prior	Project records

Source: Authors' 1997 survey data.
Note: Primary data for 1987 are based on respondents' recall in 1997.

and one irrigated plot within each category.[13]

About 30 percent of the plots are irrigated, with the proportion of irrigated land higher on better quality land (Table 3.8). (About 21 percent of plots were irrigated at the start of the study period in 1987.) Sixty-two percent of class II lands are irrigated, while the corresponding figures are 37 percent for class III lands and 7 percent for class IV lands.[14] This distribution arises for two reasons. First, given the choice, farmers will irrigate their better lands first, since water will give higher returns when applied to better soil. Second, land class is somewhat endogenous with respect to irrigation, since irrigated plots tend to be leveled and receive higher organic matter inputs. As a result, it is possible for a given plot to change to a higher land classification after it becomes irrigated.

Sampled plots were not stratified by the farmer's total landholding size, but the sample is distributed fairly evenly across respondents of different landholding size (Table 3.9). It is important to note that the sampling approach oversampled large plots relative to small ones, because the transect line was more likely to cross a large plot than a small one. However, plots are not large overall; the mean size is 0.72 hectare and the median is 0.42 hectare.

The sampling approach led to a reasonably even distribution of plots across categories when classified by soil, irrigation, and size of landholding. Other factors were not controlled, however; for example, village and household characteristics that may affect productivity and natural resource conditions at the plot level may vary across project category. Data were collected to incorporate these factors in the plot-level analysis. The various characteristics to be examined in the plot-level analysis and the source of data used for this study are found in Table 3.10.

Characteristics of Household and Interest Group Respondents

In addition to demographic and socioeconomic data collected for the household farming each sampled plot, a detailed household survey was conducted to learn about how watershed projects affected welfare and other factors potentially influencing development. The 347 respondents—sampled randomly and stratified by landholding size—differed from those in the plot survey in that about 20 percent of them were entirely landless; the remaining respondents also had smaller holdings on average than those in the plot survey. This report draws on the household survey for only a limited amount of tabular analysis, so it is not described in detail here.

Group interviews to collect qualitative data were collected in the same 29 villages as the plot data. The respondents for these interviews included the village's elected leader, or *sarpanch*, representatives of the watershed agency, and specific interest groups in the village such as farmers with irrigated land, farmers without irrigation, landless people (often herders), people from scheduled castes and tribes, and others. Men and women were interviewed in separate groups. Facilitators of these discussions had a list of unstructured questions to ask, but they also encouraged participants to address other issues of importance to them.

---

[13]In some villages, the sampling approach was altered somewhat because there were not enough plots with the desired irrigation status or land capability classification. For example, in some villages land quality could only be divided into two categories, and in others there were no irrigated plots on land of below-average quality.

[14]Since the standard land classification system goes from I (ideal land) to IX (mostly rock), the class II and III lands in the area of this study were adequate for cultivation, whereas the class IV lands were barely arable. Most of the uncultivated lands in the sample fell in class VI or VII.

CHAPTER 4

Methods

This chapter discusses the approach taken to identify the contribution of watershed projects to agricultural productivity, natural resource management, and poverty alleviation, taking into account the contributions of other factors such as infrastructure development, agroclimatic conditions, and village-level social capital. The analysis is structured around the basic program evaluation question: what would have been the state of agricultural productivity, natural resource management, and poverty in the absence of the project interventions? Answering this question is complicated because, obviously, one can never observe the same villages (or households or plots) participating and not participating in the program at the same time. As a result, one must be careful in identifying other factors that may have contributed to observed outcomes. These include contemporaneous events, such as changes in infrastructure and market access, and systematic biases introduced by the choice of villages or the people who choose to participate. Since projects are not placed randomly, differences in project outcomes may depend on preexisting village characteristics in addition to project activities.

Historically, researchers who used quantitative methods of evaluation saw no value in qualitative approaches and vice versa. Recent years, however, have seen a growing appreciation of the benefits of combining the two approaches (Greene and Caracelli 1997; Patton 1997). Quantitative evaluation uses statistical analysis to disentangle project effects from intervening factors, relying mainly on theory to explain how the project activities lead to impact. It follows the logical positivist belief that a single, objective truth exists independently of the observer. Qualitative evaluation, on the other hand, tends to make fewer assumptions about how a project affects individual behavior. It focuses on the mechanisms of change, while also yielding qualitative measures of impact. Combining the two types of information can yield a particularly thorough understanding of project impact.

This study uses mainly quantitative analysis, but it also draws on qualitative information to better understand the relevant research questions and to identify projects' unintended consequences and the mechanisms through which they operate. This sets the stage for a better-informed quantitative investigation, with greater confidence that statistical analysis addresses the most important questions and incorporates the most relevant variables. Subsequent qualitative investigation then helps to interpret the findings of the statistical analysis and to rule out competing explanations for observed differences across projects. This is particularly important given limitations in the data.

Assessing Endogeneity in Program Placement

The criterion by which each project selects participating villages is of critical importance to the present analysis. If, as argued above, numerous factors can determine a village's performance in agricultural production and natural resource management, then it is important to know how these factors are distributed across villages in different project categories. Otherwise, if villages in different project categories vary in their endowment of factors that can affect performance, then it is difficult to know whether to attribute differences in performance to project activities or to the effects of preexisting village characteristics. For example, Pitt, Rosenzweig, and Gibbons (1993) describe a case in Indonesia that showed that villages covered for several years under a major family planning program actually had higher fertility rates than those outside of the program. One could jump to the conclusion that the family planning program had failed miserably, but Pitt, Rosenzweig, and Gibbons explain that the difference was not surprising given that the program consciously worked in villages where fertility had been higher to begin with. In the absence of the family planning program, the difference in fertility between the two sets of villages might have been even greater. An analogous situation could apply in the present study, since programs may choose to operate in particularly favorable or unfavorable villages, either intentionally or unintentionally.

The critical problem in quantitative evaluation is endogeneity, which arises if some factors affect the project placement and the outcome simultaneously. This makes it difficult to distinguish the effect of the project from the underlying factors that determined where the project operated. The only way to solve this problem completely would be to observe the same individual at the same point in time, both with and without the project. Of course this is not possible, so the evaluator must try to set up one group of observations affected by the project treatment, and a control group not affected by the project but identical in every other way. This follows the standard experimental design of the natural scientist. In practice, in many social science settings it may be possible to identify *similar* but not *identical* treatment and control groups, so the social scientist's experimental design can never be perfect (Manski 1995). This is discussed later in this report with respect to watershed projects.

Evaluators follow three main approaches to establishing control and treatment groups: randomization, or pure experimental design; quasi-experimental design, and nonexperimental design (Ezemenari, Rudqvist, and Subbarao 1999). Randomization refers to randomly placing individuals into two groups—one that receives the project treatment and one that does not. This solves the endogeneity problem by ensuring that the two groups are statistically equivalent, so that any difference in average outcomes after the project can be attributed to the project. In the present context, five separate randomly placed groups would be needed: four for the different watershed projects and one control. Obviously the randomization must take place before the projects begin. In the case of Indian watersheds, the different projects operated independently and little or no advance thought was given to evaluation, so the randomization approach is not possible.

Quasi-experimental design involves matching program participants with a comparable group of individuals who did not participate in the program. This simulates randomization but need not take place prior to the intervention. For example, Pitt and Khandker (1996) used such an approach to estimate the effects of microcredit programs in Bangladesh. They matched program and nonprogram villages and then devised an elaborate set of matched groups in both sets of villages based on eligibility requirements for participating in the programs. Jalan and Ravallion (1998) used a statistical technique called propensity matching to match on the basis of multiple factors. This involves modeling the probability that each site participates in a project as a function of all observable variables known to affect participation, and then matching pairs of participating and nonparticipating sites that have an equal probability of having been selected for the project. Project impact is estimated as the mean of the differences between all matched pairs on the outcome variable.

In the present study, villages in each project category were matched geographically, but there was insufficient data to match them in a more rigorous manner. Propensity matching was not an option, for example, because it was impossible to find sufficient cases that had an equal probability of being selected for each of the five project categories. Other quasi-experimental designs were also considered. For example, one feature of Pitt and Khandker's (1996) experimental design was that the survey included both participating and nonparticipating households in villages where projects were active, as well as households from nonproject villages. That approach does not work for a study of watersheds, however, because the project may affect everyone in each village—there is no such thing as a nonparticipating household. Yet another quasi-experimental design considered is the regression-discontinuity design, whereby treatment and control groups are distinguished on the basis of a threshold score on some preproject measure (Trochim 2000). In this study, the idea would be to find a continuous variable whose values for nonproject villages and project villages tended to fall on opposite sides of some threshold value. No such measure could be identified, however, so the method could not be used. This method is also better suited to problems with only one program treatment rather than the four different programs in this study.

For this study a nonexperimental design was used instead. Several nonexperimental approaches are possible. One comes from the Pitt, Rosenzweig, and Gibbons (1993) study of Indonesian poverty programs referred to in Chapter 1. Because those programs were intentionally located in the poorest areas, purely cross-sectional analysis would have suggested mistakenly that the programs actually increased poverty. More formally, the simple approach that yielded the incorrect finding was as follows:

$$Y = a + bW + cX + e, \qquad (1)$$

where the outcome (Y) is a function of the watershed project treatment effect (W) and other determining factors (X), a is the intercept, and e is the error term. This approach is valid if the other factors (X) include all possible determinants of treatment effect and if program placement is independent of treatment effect. Since this is unlikely to be the case, the estimated effect of the treatment is likely to be biased. In the case analyzed by Pitt, Rosenzweig, and Gibbons (1993), the project coefficient b in equation 1 had a negative coefficient; that is, the analysis suggested incorrectly that the project contributed to poverty.

Pitt, Rosenzweig, and Gibbons approached this problem by estimating the change between poverty conditions before and after the project as a function of the *change* in hypothesized determining factors, such as demographic conditions and access to services and markets. In this model, the nonproject determining factors are grouped into socioeconomic variables (X) that vary both spatially and temporally, and environmental variables (E) that vary across villages but are fixed over time. The relationship can be expressed as[15]

$$\Delta Y = a\Delta W + b\Delta X + cE + \Delta e. \qquad (2)$$

This approach isolates the changes associated with the project, eliminating the bias associated with the influence of preexisting conditions on both the program placement and the outcome. However, it presumes availability of panel data (containing conditions both before and after the program), and it presumes that sufficient change occurred in the socioeconomic variables X to estimate them. Over the 10-year period of the present study many socioeconomic variables in question, such as infrastructure conditions and the distance to markets and services, did not change in most villages. As a result, many variables contained mainly values of zero, with insufficient variation within the sample to perform econometric analysis. Wherever possible, variables are expressed in terms of the change during the project period, but for most explanatory variables the value at the start of the project period is used.

Another way to control for endogeneity of program placement in estimating program effects is through instrumental variables. A variety of two-stage models for estimating treatment effects or sample selection bias provide models for this approach (Maddala 1983; Greene 1990). One equation yields the predicted probability that any given case is selected (or self-selects) for treatment under a given program. Then, in a two-stage model, another regression estimates the outcome in question, replacing the endogenous treatment variable W with its predicted value C, eliminating the endogeneity. In this case the model is as follows:

$$W = a + bX + cZ + e_1, \text{ and} \qquad (3)$$

$$Y = f + gC + hX + e_2, \qquad (4)$$

where X is a set of variables correlated to both the outcome Y and the placement of project treatment W, Z is a set of variables that affect W but not Y, and e_1 and e_2 are error terms.

In the present context, equations (3) and (4) can be written more specifically as

$$W = a + bV + cZ + e_1 \text{ and} \qquad (5)$$

$$Y = f + gC + hV + iH + jP + e_2, \qquad (6)$$

where W is a categorical variable indicating one of five watershed project categories; C is the predicted probability that the project falls in each watershed project category; Y represents outcomes defined in terms of the performance indicators introduced in Chapter 3; V is a set of village-level explanatory variables; H is a set of household-level variables; and P is a set of plot-level explanatory variables. Both H and P are omitted from village-level analysis.

[15]In this specification W may be considered as program expenditure in a given location, so W is the change in program expenditure between two points of time. W could also be specified as project dummy variables, in which case W would replace W in equation (2).

Equation 5 is a multinomial logit model because *W* is categorical. Equation 6 takes different forms depending on the nature of the performance indicator in question; these variables may be continuous, binary, or ordinal. In most of the models equation 6 is an ordinal logit model, in some it is a binary probit, and in a few it is a tobit or an ordinary least squares regression. In all of these cases, the models are adjusted for the use of complex survey data with stratification, sampling weights, and clustering (Stata Corporation 1999). These adjustments tend to cause larger standard errors in regression analysis and provide an acceptable correction for choice-based sampling.[16]

A remaining shortcoming of the model is that, for technical reasons, the standard errors could not be corrected for the fact that predicted values were used in the regressions. The author is not aware of formulas to correct the standard errors for the complex two-stage regressions used in the analysis. Bootstrapping could not be justified due to the small number of observations per stratum. Pender and Scherr (1998) faced this same problem; they examined the robustness of their findings by comparing their regression results using actual versus predicted values. This study follows the same approach, and the findings are discussed below.[17]

Qualitative Analysis

Qualitative investigation took the form of detailed, open-ended discussions, mostly at the group-level with people from different interest groups. The findings from this work helped identify some of the questions posed in the quantitative analysis, and it also helped interpret the findings. This study was primarily quantitative, so the qualitative data played mainly a supporting role. In a few cases data limitations prevented the quantitative analysis from yielding any useful information, so the qualitative analysis became the sole source of insight from the fieldwork. However, time constraints limited the scope of the qualitative investigation to less than would be ideal. In particular, it would have been desirable to engage in a more thorough qualitative investigation after having analyzed the quantitative data.

Qualitative data were recorded in written notes and yielded a variety of forms of data. Some findings from these interviews could be translated into numeric data, while others helped to explain the specific problems that people faced or the ways that projects affected them. Findings from these interviews are presented in this report alongside those from quantitative analysis as a means of providing additional insight.

[16]Choice-based sampling occurs when one category of interest has to be oversampled in order to make up for its rare occurrence in the population. Such is the case in this study for many of the project categories. Stata's survey commands weight each observation by the inverse of the probability that it is drawn, so that oversampled categories receive less weight in the analysis. They also adjust the standard errors to account for the fact that clustered observations (such as households within a village) are not independent, and that observations across different strata (for example, in separate projects) are independent. These adjustments impose strict enough restrictions on the analysis to serve as a practical tool to account for choice-based sampling (Jeffrey Wooldridge, Michigan State University Department of Economics, personal communication.)

[17]Using predicted values to instrument for program placement endogeneity can potentially cause an identification problem that can change regression results (Morduch 1999). The author is not aware of methods to correct for this problem in the complex, two-stage regressions used in this analysis. As mentioned above, the possibility of major changes in results is examined here by comparing the regression results using actual and predicted values.

CHAPTER 5

How Projects Choose Where to Operate

The way a project selects villages in which to work can have a major impact on what it can achieve, since a number of conditioning factors can strongly influence people's incentives to invest in land improvement. This chapter reviews each project's published site selection criteria and then examines the data to characterize villages under each project.

Project Site-Selection Guidelines

Before 1995, the *DPAP* focused on small microwatersheds located within predominantly rainfed villages with relatively little irrigated area. The irrigation threshold varied depending on the village's average annual rainfall; in villages with more than 1,125 millimeters average annual rainfall, the DPAP could operate if irrigated area was less than 10 percent, whereas if rainfall was less than 750 millimeters, irrigated area could be up to 20 percent.

The *Jal Sandharan* project selected villages according to four conditions.[18] First, they would be selected if COWDEP had operated there and completed more than 50 percent of the watershed development work. This would enable treatment of the entire watershed to be completed through the Jal Sandharan program within five years. Second, they should have a scarcity of drinking water. Third, they should be located in a *taluka* (a subdistrict administrative unit containing up to 200 villages) with scarce groundwater (as designated by the state Groundwater Survey and Development Agency). And fourth, they must lie outside of the command of a canal irrigation project.

Unlike the other projects in the study that work in one or two villages at a time, the *World Bank Pilot Project* worked in very large, contiguous areas of about 25,000 hectares with at least 750 millimeters average annual rainfall and little irrigation. Villages in the core watershed area were treated in their entirety, whereas those in the periphery typically lie partly in the watershed and partly outside. In that case only the part of the village lying inside the watershed was treated.

The *NWDPRA* operated in areas with less than 30 percent irrigation, with no criteria concerning average annual rainfall. Preferably sites should be located in the upper reaches of the local macrowatershed. Project sites should be close to the *taluka* or block headquarters in order to facilitate supervision by officials based at headquarters, and close to markets so that "farmers from nearby areas can assemble and see the process and feel the impact of the project interventions (India, Ministry of Agriculture1991)." Project sites should be located on the main road, easily accessible to government officials and other visitors. To quote the

[18]Where DPAP funds are used, its criteria are also followed.

guidelines, "Just a pause on the road would give an opportunity to have a bird's eye view of the project area. This will ensure visual impact on intentional and unintentional visitors (India, Ministry of Agriculture 1991)."

Each *NGO* has its own guidelines, but virtually all of them stress working in poverty-stricken areas, often inhabited by tribal groups. MYRADA's guidelines are instructive. It operates in remote, unfavorable areas, usually in the border areas of a state, far from the state capital, which are relatively neglected by state-level development programs. They have the worst land, the worst infrastructure, and the least well-off inhabitants. The typical MYRADA village has very different characteristics from those in the NWDPRA project sites described above.

MYRADA also avoids watershed development where more than 10 percent of the population is landless for fear that it would be impossible to generate sufficient benefits for them (Fernandez 1994). Protecting livelihoods while restricting access to common lands might not be feasible. In recognition of the potential equity tradeoffs involved with watershed development, MYRADA undertakes other kinds of development activities in such areas.

Any village is eligible for participation in the *AGY* if it is located in a drought-prone area of Maharashtra, with no more than 30 percent cultivated area under irrigation. The villagers must meet in the Gram Sabha (assembly of all adults in the village) to pledge to accept and abide by the five social principles listed above. Seventy percent of the Gram Sabha must agree to participate before the application can go through.

Villages selected for an *IGWDP* project should be in a drought-prone area with less than 20 percent area under irrigation and overall water scarcity; they should lie in the upper part of a macrowatershed and have noticeable erosion, land degradation, and resource depletion problems. Village boundaries should coincide with watershed boundaries to the greatest extent possible, and the topography should offer good opportunities for water harvesting. Villages should be predominantly poor with a high proportion of scheduled castes and scheduled tribes in the population, and landholdings within selected villages should be relatively equally distributed. Beyond this, the villagers should commit to the social conditions outlined. The key factor is that villagers should demonstrate their capacity for collective action and their concern for resource conservation. Finally, the village is represented by a village watershed committee that is selected on the basis of consensus in the Gram Sabha (NABARD 1995; Farrington and Lobo 1997).

The site selection criteria reveal a great deal about each project's orientation. All of the projects share a bias toward working in areas that are less favored agroclimatically, although it is stronger in some than others. The NWDPRA is most lenient about agroclimatic conditions because of its interest in developing rainfed agriculture.

The NWDPRA is also the only project that does not seek to work in villages that are least well off socioeconomically. Selecting the most easily accessible villages reveals the NWDPRA's orientation toward planning and supervision by people from outside the village, as well as an optimistic view about the process of dissemination of project benefits. More subtly, the approach also leads to an apparently unintentional bias in selection of project sites toward more densely populated areas with better access to transport and markets. In accordance with the conceptual framework outlined in the introductory chapter, these conditions may well be especially favorable for the promotion of rainfed areas. In this sense the project's technical interventions may complement other features of the project sites.

The focus on social discipline is greatest in AGY and IGWDP. In many respects the villages under this project are self-selected for collective action. Many of them practiced the required social restrictions prior to the onset of the program. Villages that are not prepared to ban grazing and tree cutting or to practice *shramdan* shy away from these programs. The IGWDP takes the additional step of selecting only villages with topography favorable for constructing water-harvesting structures.

Another point about selecting villages for success is that in the start-up phase of the IGWDP, which is covered under this study, only well-established NGOs were selected—NGOS already familiar with the community in which the project was to be initiated. That means that they were already working there prior to the start of the IGWDP, and many ongoing activities were simply brought under the flag of the IGWDP. A similar situation holds for the AGY. Many AGY project villages are led by disciples of Anna Hazare and were attempting to follow his development philosophy even before the project began.

Analysis of Determinants of Program Placement

This investigation of how published site-selection guidelines translate into actual program placement in the study villages focuses on Maharashtra because the data in Andhra Pradesh were limited, as discussed earlier.

Table 5.1 presents mean values of variables that describe the villages in Maharashtra covered by the village survey. The figures are for 1987, before the projects began. The data suggest that the projects do in fact follow the principles laid out in their guidelines. NGO and NGO/government collaborative projects tended to have the lowest levels of infrastructure before the project began, while NWDPRA villages had the highest. Conditions in villages with no projects were similar to those under the NWDPRA. All of the NGO/government collaborative project villages practiced *shramdan*, as did 75 percent of the NGO villages. Less than half of the remaining villages practiced *shramdan*. None of the projects had a significantly lower percentage of area irrigated than the nonproject villages, but this is because virtually all villages in the study area had relatively little irrigation, and also because the government projects targeted their work based on the level of irrigation at the *taluka* level, not the village level.

As mentioned, nearly all projects took advantage of work done by the COWDEP project in the 1980s, although only Jal Sandharan advertised this fact in its published guidelines. Only three project villages in Maharashtra—one each under the NWDPRA, DPAP, and NGO categories—were not previously treated under COWDEP. For NGOs, selecting COWDEP villages is sensible because technical work that was already undertaken can be made more productive by developing complementary social institutions. At the same time, it makes it somewhat difficult to determine how much of the success of the program stems from the project's current work and how much depends on earlier watershed structures built under COWDEP. The time frame required for watershed development interventions may change when the project expands to areas beyond those covered by COWDEP.

Another interesting observation in Table 5.1, not revealed by published guidelines, is that in all the project categories, a smaller percentage of villages contained common land (owned by the government) than in nonproject villages. The difference was particularly large in NGO villages; only 33 percent contained government revenue land, compared with 71percent for nonproject villages and 57 percent overall. Such land usually lies in the upper watershed and is used as grazing commons. As discussed in Chapter 1, watershed projects aim to restrict access by grazing animals to this land. Throughout India, open-access government revenue land is in a notoriously degraded state. Accordingly, the organizational requirements of watershed development may

Table 5.1 Mean values of selected Maharashtra village characteristics in 1987, by project category

Village characteristic	All	NWDPRA	JS	NGO	AGY or IGWDP	No project
Irrigated area (%)	13.0	24.0	13.0	9.0	16.0	11.0
Villages located in the macrowatershed (%)						
Upper reaches	59.0	70.0	47.0	75.0	64.0	47.0
Middle reaches	23.0	20.0	41.0	17.0	0.0	29.0
Lower reaches	19.0	10.0	12.0	8.0	36.0	24.0
% of villages located on a:						
Highway	4.0	0.0	12.0	0.0	7.0	0.0
Paved road	34.0	50.0	29.0	42.0	21.0	35.0
Good unpaved road	24.0	30.0	18.0	17.0	43.0	18.0
Bad unpaved road	29.0	20.0	35.0	25.0	14.0	41.0
Bullock cart path	9.0	0.0	6.0	17.0	14.0	6.0
Villages with a public health service office (%)	16.0	50.0	20.0	10.0	0.0	10.0
Houses with an electrical connection (%)	27.0	32.0	25.0	24.0	21.0	32.0
Villages with electricity to power irrigation pumps (%)	89.0	100.0	88.0	92.0	78.0	88.0
Villages with informal credit groups (%)	24.0	60.0	12.0	25.0	0.0	35.0
Villages with adequate drinking water (not delivered by tanker) (%)	53.0	70.0	35.0	50.0	43.0	71.0
Villages containing some common (government revenue) land (%)	57.0	60.0	59.0	33.0	57.0	71.0
Villages containing some Forest Department land (%)	76.0	80.0	71.0	83.0	86.0	65.0
Villages practicing *shramdam* (%)[a]	54.0	20.0	35.0	75.0	100.0	41.0
Villages obtaining resources from government land (%)						
Fuel[b]	40.0	33.0	40.0	50.0	63.0	25.0
Grass fodder[b]	83.0	83.0	90.0	50.0	88.0	83.0
Tree fodder[b]	40.0	50.0	40.0	50.0	50.0	25.0
Literacy rate (1991 census) (%)						
Overall adult	53.0	58.0	54.0	46.0	52.0	56.0
Male	68.0	73.0	69.0	62.0	67.0	73.0
Female	37.0	43.0	37.0	30.0	38.0	40.0
Distance in kilometers to nearest:						
Regulated market[c]	11.7	13.9	11.6	16.8	8.1	9.8
Taluka headquarters	20.6	18.4	18.9	24.7	22.1	19.5
District headquarters	65.6	60.8	61.6	65.4	69.1	69.7
Large city[d]	77.8	56.8	94.7	88.3	65.8	75.7
Bus stop	1.2	0.9	0.9	1.7	1.4	1.3
Industrial unit	24.0	19.8	25.4	28.7	26.8	20.7
Bank[e]	6.0	3.5	5.7	9.2	6.7	4.9
Public health service	6.9	3.7	7.2	7.5	9.4	6.0
Veterinarian	5.4	3.2	5.5	5.9	6.9	4.9

(continued)

Table 5.1—continued

Village characteristic	All	NWDPRA	JS	NGO	AGY or IGWDP	No project
Average annual rainfall (millimeters)	578.0	601.0	566.0	593.0	583.0	578.0
Altitude range between highest and lowest point in the village (meters)	49.0	67.0	54.0	51.0	44.0	37.0
Population density (inhabitants/square kilometer, 1991 census)	135.0	162.0	123.0	118.0	131.0	145.0
Number of extension agent visits per month	1.6	1.4	1.5	1.3	2.2	1.6

Source: Authors' 1997 survey data.

Notes: Analysis of variance statistics are reported for continuous variables with significance level $p < .10$. Kruskal-Wallis statistics for ordinal categorical variables are reported for variables with significance level $p < .10$ (Agresti and Finlay 1997). See Chapter 2 or the glossary at the end of the report for full names and descriptions of the projects.

[a]Whether *shramdan* (community voluntary labor) is used: chi-square = 21.9, 4 df, $p < .001$.

[b]These measurements cover only those villages that had government revenue land in both 1987 and 1997.

[c]Distance to nearest regulated market: $F = 2.80$, 4 df, $p < .03$.

[d]Distance to nearest large city: $F = 2.84$, 4 df, $p > .05$.

[e]Distance to nearest bank: $F = 2.07$, 4 df, $p < .1$.

be significantly less complicated in villages without government revenue land. In addition, working where there are no common lands may reflect the concerns expressed by Fernandez (1994) that watershed development cannot support the livelihoods of large numbers of landless people. Where there are no common lands there is less potential for negative effects of watershed development on landless people.

Another kind of common land is owned by the Forest Department, which uses its own resources to restrict access. More project villages than nonproject villages contain Forest Department land, probably because they are likely to be located in hilly terrain with more forested area.

Challenges in Selecting "Control" Villages

It is important to note the difficulty—or perhaps impossibility—of selecting nonproject villages in a way that provides a true "control" against which to measure the performance of the watershed projects.

If it were possible to sample enough villages to select those with identical probability of falling in each of the five categories, one could be quite confident in the adequacy of the control category. Jalan and Ravallion (1998) used this propensity matching technique in their work in Argentina, for example. In the present study, obtaining a sample in which every village was matched by villages from other project categories, all with identical probability of falling in five separate project categories, may have been impossible. At the very least, even coming close to this ideal would require a very large sample size to locate the villages with the required probabilities. Data collection for this procedure would require more resources than were available for this study.

As a result, it is inevitable that there will be some degree of systematic difference across project categories and there will not be a pure control in the experimental science sense. A quick scan of Table 5.1 supports this notion: nonproject villages never have the same mean characteristics as other project categories. In most cases the mean value for nonproject villages lies within the range of other project categories, but in some cases it lies on the extreme. For example, nonproject villages are on average flatter than other villages with a mean altitude range of 37 meters, whereas the mean altitude range for other categories varies from 44 to 67 meters. Similarly, nonproject villages are more likely to contain common land as mentioned above. The differences are not statistically significant, but it is clear that the nonproject villages are not identical to the other categories. This supports the need to utilize the instrumental variables analysis described in Chapter 4.

Econometric Analysis

For the econometric analysis of the determinants of project placement, a multinomial logit model is used to examine in more detail the determinants of which project category a particular village falls into. The dependent variable is the categorical project variable covering the five categories found in Maharashtra: NWDPRA, Jal Sandharan, NGOs, and AGY/IGWDP (combined into one category), and nonproject villages.

Explanatory Variables. Conditions prevailing in 1987, before the projects began, represent the potential determinants of a village's selection by a given project. Altitude range (the difference between the highest and lowest points, in meters) is important since many projects seek to work in areas with high potential for water harvesting. Infrastructure variables include the distance to *taluka* headquarters, the population density in 1990,[19] percent of area irrigated, adequacy of drinking water availability, distance to market, distance to the nearest bus stop, and distance to the nearest public health center. Other infrastructure variables are omitted because they were highly correlated with those included.[20] The existence of an old COWDEP project in the village is omitted from the analysis because it perfectly predicts the existence of a current project, making the multinomial logit analysis infeasible. Variables representing social conditions and social institutions include whether the village practiced *shramdan* and the number of communal groups. The literacy rate was considered but excluded because it is highly correlated with some of the infrastructure variables. A dummy variable representing the presence or absence of common land is also included.

Results. The multinomial logit analysis supports most of the descriptive findings about project selection and raises some additional points (Table 5.2). With nonproject villages as the base category, the analysis shows the following results.

All projects have a greater range in altitude between the highest and lowest point in the village, compared with nonproject villages, and this difference is significant for all except the NGO/government collaborative projects. This is to be expected since hilly areas are most suited for water harvesting.

The AGY and IGWDP villages were significantly more likely to practice *shramdan* in 1987. NWDPRA villages were actually significantly less likely to practice *shramdan* than nonproject villages; the reasons for this finding are not known. NWDPRA, Jal Sandharan, and NGO villages all

[19]Population data come from the 1990 census (India, Office of the Registrar General 1991); they are not available for 1987.

[20]An effort was made to build an index of infrastructure quality, but it had limited explanatory power.

had more communal diversity and more people of scheduled castes and tribes and backward classes than nonproject villages, and the latter is consistent with published guidelines. AGY and IGWDP, on the other hand, had no significant differences from nonproject villages in communal diversity and scheduled castes and tribes. If the analysis is conducted using the AGY and IGWDP as the base category (not shown) the communal diversity and population of scheduled castes and tribes are significantly

Table 5.2 Determinants of project category in Maharashtra (multinomial logit regression)

	Project category			
Variable	**Ministry Agriculture**	**Ministry of Rural Development**	**NGO**	**NGO/ government collaboration**
Distance to nearest bus stop in 1987 (kilometers)	0.83	-0.16	0.16	-0.34
	(.34)**	(0.27)	(0.32)	(0.29)
Paved road in 1987 (dummy)	0.29	-1.58	0.41	-2.49
	(1.27)	(1.63)	(1.11)	(1.53)
Whether the village contained government revenue land, 1987	-0.32	-2.10	-4.96	-1.16
	(1.16)	(1.22)*	(1-17)***	(0.88)
Number of communal groups in the village	1.18	0.76	0.85	0.13
	(0.25)***	(0.29)**	(0.30)***	(0.35)
Altitude range (hundreds of meters)	3.34	1.93	2.44	2.16
	(1.02)***	(1.00)*	(1.06)	(1.34)
Distance to *taluka* headquarters (kilometers)	0.21	0.01	0.35	-0.03
	(0.05)***	(0.05)	(0.43)	(0.04)
Population density in 1990 (hundreds of persons/square kilometer)	3.71	0.88	-1.81	-0.59
	(0.82)***	(1.76)	(-1.43)	(0.88)
% area irrigated in 1987	2.90	-2.39	8.29	1.94
	(3.28)	(5.76)	(3.55)**	(4.52)
Whether the village had sufficient drinking water in 1987 (dummy)	3.31	-1.35	0.26	0.93
	(1.38)	(1.27)	(1.54)	(1.49)
Distance to nearest public health center, 1987 (kilometers)	-0.38	0.17	0.18	0.33
	(0.15)**	(0.15)	(0.15)	(0.14)**
Distance to market for agricultural inputs in 1987 (kilometers)	-0.15	0.23	0.34	0.10
	(0.11)	(0.15)	(0.16)**	(0.13)
Village practiced community voluntary labor in 1987 (dummy)	-2.01	-1.31	1.57	8.42
	(1.10)*	(1.51)	(1.57)	(2.35)***
Area of the village (hundreds of hectares)	0.17	1.29	0.09	0.30
	(0.13)	(1.34)	(0.13)	(0.13)**
% of households with at least one seasonal migrant, 1987	-0.10	-0.06	-0.10	0.09
	(0.03)***	(0.04)	(0.06)	(0.03)***
% of inhabitants of SC, ST, BC	0.047	0.08	0.12	-0.03
	(0.025)*	(0.03)***	(0.03)***	(0.06)

Source: Authors' 1997 survey data.

Notes: Reference category is no project; variables reflect values in the preproject period. There are 70 observations. Model is not corrected for choice-based sampling; the sample is stratified on the dependent variable. Coefficients and standard errors are adjusted to account for sampling weights, stratifications, and finite population size. SC is scheduled castes, ST is scheduled tribes, and BC is backward classes.

*Statistically significant at 10 percent.

**Statistically significant at 5 percent.

***Statistically significant at 1 percent.

lower than other project categories. The IGWDP requires consensus-based decision-making, which may be easier with communal homogeneity, and the two projects require a ban on open grazing and tree cutting, which may be more difficult for poor, landless or near-landless people to accept because they rely on products from the commons for their livelihoods. NGO and Jal Sandharan villages were significantly less likely to contain government revenue land, possibly suggesting that these projects sought to reduce the potential for equity tradeoffs. NWDPRA and AGY/IGWDP villages also were less likely to contain government revenue land, but the difference is not statistically significant.

NWDPRA villages were likely to have significantly fewer seasonal migrant workers, while AGY and IGWDP villages were likely to have significantly more. A higher number of migrant workers can be an indication of poor economic conditions locally. AGY and IGWDP villages are also significantly larger in area; the reason for this difference is not clear.

NWDPRA villages were likely to be more densely populated and other villages less densely populated than nonproject villages, but this difference is significant only for the NWDPRA villages. This is consistent with the NWDPRA's published guidelines, which call for working in more accessible, visible villages. Again, this probably reflects a nonrandom selection process. NWDPRA villages are also closer to public health clinics and markets, though only the former is significant. NGO villages, on the other hand, were significantly likely to be located further from markets and *taluka* headquarters, and AGY and IGWDP villages were significantly farther from the nearest public health office.

Finally, only Jal Sandharan and DPAP villages were more likely to have a drinking water shortage, consistent with the project's mandate, but the difference was not statistically significant.

CHAPTER 6

Natural Resource Management and Productivity on Uncultivated Lands

Most nonarable lands in the study region are managed either as government revenue land or Forest Department land. While the Forest Department heavily restricts access to its lands (at least in principle), access to government revenue land is typically open to all users. Protecting it requires village-level management institutions based on widespread commitment to improvement of this resource.

Watershed projects seek to develop nonarable lands for a variety of reasons. In projects oriented toward water harvesting, nonarable lands are typically in the upper reaches of the watershed, which act as the catchment area for water-harvesting structures downstream. If the upper reaches are poorly maintained, erosion will silt the water-harvesting structures, rendering them useless. So developing and protecting nonarable lands is a prerequisite to the primary objective of raising the water table.

Developing nonarable lands also has direct benefits, particularly if it increases the long-term availability of products such as fuel and fodder historically supplied by these lands. Soil and water conservation trenches are dug to concentrate water and soil, with trees and grasses planted in the trenches. In the early years after planting, the common lands must be strictly protected against grazing so that plants can become established. After that, they can supply a steady stream of fodder and fuel as long as grazing and harvesting are restricted.

An essential feature of revegetation measures is the need to leave undisturbed the recently planted material. Revegetated areas are off limits to grazing for a prescribed period, ranging from three to seven years, depending on growing conditions. During the first year or two, people are encouraged to leave the area completely untouched during the growing season and harvest it by hand after the season. Once some regeneration has taken place they may be allowed to begin controlled harvesting during the rainy season. Grazing remains off limits, however, until the new vegetation is well established.

This system imposes obvious hardships on those who depend on the uncultivated lands for their livelihoods, including herders who use the land for grazing and women who collect fuel from it. People with no land of their own are particularly affected. Projects try to compensate them for their losses by putting people to work planting vegetation and installing conservation structures. Eventually they are expected to benefit from the rehabilitated common land. For the first few years, however, they lose an important component of their livelihood.

The typical scenario on the common lands in rural India has been one of gradually declining productivity as a result of overexploitation, which in turn resulted from institutional arrangements that were inadequate to encourage people to protect and develop these lands (Singh 1997). Historically, management of common lands followed at least three different patterns (Gadgil and Guha 1992). In some places they were accessible to all, with insufficient pressure on resources to lead to severe degradation until the last several decades. In others, management was enforced by powerful landowners, who acted as gatekeepers to make sure that the common lands were not overexploited (Gadgil and Guha 1992; Bentley 1984). While this system was good for the condition of the land, it was inequitable because the landlords received most of the benefits. In a third kind of situation, democratic village-level institutions resulted in sophisticated, equitable ways of sharing both rights and responsibilities for managing common lands (Agarwal and Narain 1989). Although this latter situation is sometimes presented as the historical norm in rural India, there is little evidence that it prevailed beyond a small number of villages. It is often said anecdotally, however, that such systems are still common in tribal areas.

What the Projects Do

The idea behind most current watershed project efforts on common lands is to use a combination of technical and institutional means to move the supply of products such as fuel and fodder from a low-level equilibrium to a higher one. In addition to installing soil and water conservation works and planting vegetation, most projects today seek to develop institutions for managing government lands based on principles of common property resource management. They typically encourage villagers to establish users' committees that are expected to develop and enforce management plans that satisfy the needs of every interest group. In short, they try to create an ideal, democratic arrangement.

As described in Chapter 2, the AGY, IGWDP, and NGO projects all directed relatively large efforts to social organization, and particularly to mechanisms to reduce pressure on common lands. The IGWDP and AGY, for example, only worked in villages that promised to ban grazing and cutting of trees. All of these projects also promoted social fencing—social mechanisms to achieve protection of the common lands. Such efforts may include encouraging everyone to comply with restrictions on the commons and devising arrangements to guard them if necessary. The NWDPRA and Jal Sandharan guidelines covered the same issues but in a more cursory way, and unlike the other agencies, they did not make it a central component of their work. Since none of their staff members had any training related to social organization, they were unlikely to undertake such work. The NWDPRA and Jal Sandharan had provisions to contract this part of the work out to NGOs, but only for a few weeks, after which social organization was expected to take care of itself.

An important feature of project investment on common lands is that it was entirely subsidized. This was the case throughout the 1990s and remains so at present. Some projects required a 10 percent in-kind contribution (in terms of donated labor), but this was more than offset by the fact that the projects paid above the market wage. At the very least, donating 10 percent of one's labor to gain a day of employment was a break-even proposition. In developing the common lands, the AGY and IGWDP, as well as some NGOs, obtained local support by tapping into the practice of *shramdan* (voluntary labor contributions to the community's welfare). Approximately 16 percent of the costs were to be contributed through *shramdan*, with half the value returned in cash to a village development fund for maintaining watershed development infrastructure.

The indicators presented in this chapter help identify the extent to which various types of watershed projects have succeeded in developing and protecting common lands. The section on common lands is divided into discussions of four sets of indicators. One type of indicator is the introduction of social fencing institutions to encourage protection of the commons, and the other three are rough measurements of natural resource conditions, including erosion and conservation status of the main drainage line, erosion status of nonarable lands, and changes in availability of fodder and fuel. The analysis of the condition of the drainage line covers the 64 Maharashtra villages that had a main drainage line, while the remaining analysis covers the 40 villages that included government revenue land, since villagers have the authority to manage these common lands as they please.

Social Fencing Institutions

The most common social fencing institutions are bans on grazing and cutting trees on common land. Many villages impose such bans, but whether they are adhered to is another question. Investigators, who only stayed in each village for two to four days, tried to determine whether grazing bans and tree-cutting bans were active or in name only.[21]

Table 6.1 lists the number and percentage of villages in each project category that banned grazing, as well as the number that actually imposed penalties on offenders of rules against grazing and tree cutting both before and during the project. A traditional penalty against illicit grazing, for example, is to impound the grazing animals in the *panchayat* (village government) office and release them only upon payment of a fee. *Panchayats* or watershed user organizations keep records of such payments, so it is not difficult to identify whether or not such punishment systems were enforced in 1997.

In both 1987 and 1997, banning grazing on the commons was the exception, not the rule (Table 6.1). Only 5 out of 40 villages (12.5 percent) had banned grazing before the project; the share rose to 35 percent after introduction of the projects. The share of villages that imposed punishments for illicit grazing were even smaller—5 percent

Table 6.1 Number and share of villages with restrictions on access to common (government revenue) lands, by project category, 1987 (preproject) and 1997

Type of restriction	NWDPRA	DPA and JAL Sandharan	NGO	AGY or IGWDP	No project	Total
Open grazing restricted 1987						
1987	1 (17)	2 (20)	0 (0)	0 (0)	2 (17)	5 (13)
1997	2 (33)	4 (40)	1 (25)	4 (50)	3 (25)	14 (35)
Introduced by project	1 (17)	2 (20)	1 (25)	4 (50)	1 (8)	9 (23)
Punishment for open grazing						
1987	0 (0)	1(10)	0 (0)	0 (0)	1 (8)	2 (5)
1997	1 (17)	2 (20)	0 (0)	4 (50)	2 (17)	9 (23)
Introduced by project	1 (17)	1 (10)	0 (0)	4 (50)	1 (8)	7 (18)
Punishment for cutting trees						
1987	1 (17)	1 (10)	0 (0)	0 (0)	0 (0)	2 (5)
1997	1 (17)	2 (20)	1 (25)	4 (50)	0 (0)	8 (20)
Introduced by project[a]	0 (0)	1 (10)	1 (25)	4 (50)	0 (0)	6 (15)

Source: Authors' 1997 survey data.

Notes: This table is based on the 40 Maharashtra villages that had government revenue land in both 1987 and 1997. Figures in parentheses are percentages of category totals. Category totals are NWDPRA, 6; Jal Sandharan, 10; NGO, 4; AGY or IGWDP, 8; no project, 12. See Chapter 2 or the glossary at the end of the report for full names and descriptions of the projects.

[a]Kruskall-Wallis test shows that differences across categories are significant for introduction of punishment for cutting trees (chi square = 11.1, 4 df, $p < .027$). Other differences are not statistically significant.

[21]For example, if a pasture is protected against grazing there should be no traces of cow dung on the ground.

in 1987 and 22 percent in 1997. For punishing illicit cutting of trees, the numbers are similar: 5 percent in 1987 and 20 percent in 1997. Two details in the table are particularly interesting. First, even some of the nonproject villages imposed grazing bans; clearly this is not something that necessarily requires a watershed project. Second, while none of the AGY and IGWDP villages had imposed bans or penalties in 1987, by 1997 50 percent of them had done so. They are the only category of villages with a statistically significantly higher percentage than the nonproject villages. At the same time, the 50 percent figure is low in comparison with the projects' target of universal compliance with bans on grazing and cutting trees.

No regression analysis is performed on the determinants of banning grazing and tree cutting, because so few villages actually imposed these restrictions.

Erosion and Conservation Status of the Main Drainage Line

By definition, the main drainage line is where runoff water concentrates, so it is highly vulnerable to soil erosion. The drainage line is also usually on government land, which tends to be managed poorly compared with privately operated land, so the drainage line faces management challenges from both biophysical and social causes. (Out of the 64 villages, 40 had government revenue land and most of the remainder had Forest Department land.)

Field investigators trained in soil survey methods conducted a transect of the main drainage line in each village, making visual observations of its condition and the extent of erosion on its banks. The transect was divided into segments of equal length (100 meters), and investigators made an assessment of its overall condition. This is determined by whether it appears to be under control and not expanding into adjoining fields, the extent and condition of bunds on the sides of the drainage line, and the extent of breaches in the sides of the drainage line. Each segment was assigned one of three possible scores for each of these characteristics: 3 for "good or high," 2 for "intermediate," and 1 for "poor or low." Overall scores for each village were then calculated by taking the simple average of all the segment scores. This visually based scoring system was the best that could be achieved given the resources available to the project and the lack of existing data. While it is obviously subjective, the scores should be consistent across villages within each state because only two teams of people conducted the transects, and the members rotated regularly in an effort to make sure they all used the same standards. Also, the 1 to 3 scale, while reducing the ability to make fine distinctions across observations, reduces the likely variation in scoring standards across data collection teams.

Drainage line transect scores are analyzed in both tabular and econometric form. Average values of the drainage line scores in Table 6.2 indicate that the kind of project operating in the village has a small but statistically significant effect on the drainage line, with AGY and IGWDP villages having the best average score and nonproject villages having the worst. (Note that even the best average score is only 2.00, indicating an intermediate condition.) Evidence is stronger, however, that the duration of the watershed project and the percentage of the village covered by a project have a positive effect on the condition of the drainage line, regardless of the project category (Table 6.2). This suggests that some kind of watershed activity is better than none for the condition of the drainage line.

Drainage line scores are also analyzed through multivariate econometric analysis, which is used to identify village- and project-level factors that determine the drainage line scores. Each village's score represents the mean value of scores for all the 100-meter segments of the drainage line, so the scores take continuous values ranging from 1 to 3. A Tobit model is appropriate in this case.

Table 6.2 Drainage line transect scores at the village level by project category and other factors

Village characteristic	Average score for condition of the drainage line[a]
All villages	1.70
Project category	
NWDPRA	1.78[b]
Jal Sandharan	1.60[b]
NGO	1.79[b]
AGY or IGWDP	2.00[b]
No project	1.40[b]
Total number of years under old and new watershed projects	
0 (no project)	1.40[c]
13 or less	1.60[c]
14 or 15	1.82[c]
Percentage of village covered by the project	
0 (no project)	1.40[d]
20-80	1.78[d]
100	1.78[d]

Source: Authors' 1997 survey data.

Notes: See Chapter 2 or the glossary at the end of the report for full names and descriptions of the projects.

[a]Possible drainage line scores are 1, 2, and 3. Strictly speaking, they are ordinal categorical variables, not cardinal, but average scores are shown here for ease of presentation. An average score of 2.00 means that average condition is intermediate, less than 2.00 means the condition is poor on average, and greater than 2.00 means the condition is good on average. The Kruskal-Wallis test for ordinal categorical variables shows that the following variables are significant at 10 percent.

[b]Overall condition of the drainage line varies significantly among project categories (chi-square = 8.7,4 df).

[c]Overall condition of the drainage line varies significantly by number of years under old and new projects (chi-square = 8.0, 2df).

[d]Overall condition of the drainage line varies significantly by percentage of village covered by project (chi-square = 6.0, 2df).

Explanatory variables. 1987 variables are used in the model of determinants of the condition of the drainage line, since its stabilization through soil and water conservation measures is a long-term process. 1997 values would not be the correct variables to explain the effectiveness of conservation measures that took place prior to 1997. Agroclimatic variables include the village's altitude range, which is reflected in the course of the drainage line and determines susceptibility to erosion. Average annual rainfall also determines susceptibility to erosion, but it is highly correlated to altitude range (>0.6), with more hilly areas having higher rainfall; therefore, rainfall is omitted from the model.

Social institutions and characteristics include the number of different communal (caste and religious) groups in the village, the percentage of households in the village that derive their income primarily from herding sheep, and a dummy variable indicating whether the village contains government revenue land. It is hypothesized that a higher proportion of shepherds will bring increased resistance to protecting the commons and thus poorer condition of the drainage line. The practice of *shramdan* is excluded because it is highly correlated with the predicted probability that the AGY or IGWDP operates in the village. The presence of government land almost certainly indicates that the drainage line runs through common land, which is more difficult to manage.

Economic factors include infrastructure, such as the presence or absence of a paved road, distance in kilometers to the nearest bus stop, distance in kilometers to the *taluka* headquarters, population density (inhabitants per square kilometer), and the percentage of people in the village who earn most of their income from a source other than cultivation, livestock, or agricultural labor. Population density, infrastructure, and access to markets can increase the pressure on natural resources, but they can also raise the returns to better land management. Off-farm income also has an ambiguous effect; it can help finance land improvement or it can lead people to focus their interests elsewhere, making them less willing to participate in social action to develop the village's natural resources (Gebremedhin, Pender, and Tesfaye 2000; Pender and Kerr 1998). Finally, the project inputs are represented by predicted values of the dummy variables for each project category and the average project expenditure per hectare in the village (including expenditure under the current project and previously under COWDEP).[22]

Results. The results for two cases, one in which the predicted project category is multiplied by the project expenditure per hectare and one in which these variables are specified separately, are presented in Table 6.3. There are only 64 observations because six villages have no main drainage line. The models have highly significant F-statistics, but both R^2 values are about 0.38, so the extent of variation explained by the model is not high. When expenditure and project category are specified separately (Model 1), expenditure per hectare is highly significant, but only the AGY or IGWDP variable coefficient is high among the project categories.[23] The dummy variable indicating the presence of government revenue land is positive and statistically significant, which was unexpected. The percentage of households earning their livelihoods as shepherds is negative and statistically significant, as expected. Most other variables have the expected sign but are insignificant. These include the percentage of people working outside of agriculture or livestock (–), distance to the *taluka* headquarters (–), population density (+), and existence of a paved road in 1987 (+). The latter three signs are consistent with the induced innovation hypothesis that better market access may raise the incentives to manage land better.

When the model is respecified so that project expenditure and the project category variables interact, all the project expenditure variables have positive coefficients, and all are statistically significant except the Jal Sandharan and DPAP variable, which is nearly significant. This lends support to the notion that all the projects are successful in improving the condition of the drainage line. The AGY or IGWDP category has a much higher coefficient than the other categories, as well as a higher level of statistical significance, so these projects appear to perform the best. For every rupee spent by the AGY or IGWDP, the drainage line score rises by 0.27 on a scale of 1 to 3. For NGOs it is 0.17, and for government projects the increase is less than 0.10. Other significant variables remain the same as in the previous specification of the model. A similar result is obtained with the use of actual project dummy variables, except with a smaller effect for each rupee spent. (The findings of the analysis with actual project dummy variables are not presented to save space.)

Erosion of Uncultivated Lands

Field investigators conducted a separate transect following a route perpendicular to the main drainage line, designed to cover a representative tract of the village's area, with variations in soil types, slopes, and land use.[24] The route was selected based on discussions with the *sarpanch*, groups of farmers, and a soil map of the village (where available). The investigators delineated the transect route into separate segments whose boundaries were defined by changes in either land use (nonarable, rainfed, irrigated), land capability classification, or the extent of soil erosion and soil conservation measures.

This section examines the findings regarding the extent of soil erosion on uncultivated lands in the transect, which refers to

[22]The average expenditure per hectare for the area where the project actually made investments was multiplied by the proportion of the total area in which investments were actually made. For example, if a project made investments on two-thirds of the land in a village, the expenditure was multiplied by two-thirds. Thus fairer comparisons could be made between villages with a lot of area covered by investment and those with only a little.

[23]The standard errors have not been corrected for the use of predicted values of the project category.

[24]The straight line design of the field transect oversamples plots close to the center of the village relative to those at the periphery, which are more likely to be hilly and degraded.

Table 6.3 Determinants of drainage line erosion status (interval regression)

Variable	Coefficients in model 1[a]	Coefficients in model 2[b]
Whether the village contained government revenue land in 1987 (dummy)	0.42	0.42
	(0.12)***	(0.11)***
Altitude range (thousands of meters)	-4.32	-5.85
	(8.03)	(7.88)
Distance to nearest bus stop in 1987 (kilometers)	0.04	0.04
	(0.04)	(0.04)
Paved road in 1987 (dummy)	0.15	0.17
	(0.12)	(0.12)
Population density in 1990 (hundreds of persons/square kilometer)	0.06	0.07
	(0.05)	(0.14)
Distance to *taluka* headquarters (tens of kilometers)	-0.06	-0.06
	(0.05)	(0.06)
% of inhabitants working primarily in nonagricultural sector	-0.01	-0.01
	(0.01)	(0.01)
% of inhabitants working primarily as shepherds	-0.07	-0.06
	(-0.04)	(0.04)*
Mean project expenditure per hectare (thousands of Rs)	0.06	...
	(0.02)***	
Project dummy variables for:		
NWDPRA	0.24	...
	(0.20)	
DPAP and Jal Sandharan	0.007	...
	(0.27)	
NGO	0.58	...
	(0.40)	
AGY or IGWDP	0.67	...
	(0.25)	
Mean village expenditure per hectare (thousands of Rs)		
Project operated by:		
NWDPRA	...	0.10
		(0.05)**
DPAP and Jal Sandharan	...	0.07
		(0.04)
NGO	...	0.17
		(0.06)***
AGY or IGWDP	...	0.27
		(0.05)***

Source: Authors' 1997 survey data

Notes: Observation: 64. Possible transect scores range from 1 to 3. Coefficients and standard errors are adjusted to account for sampling weights, stratification, and finite population size. Predicted values based on the multinomial logit regression in Table 5.2 are used for the four project category variables. Standard errors are not adjusted for use of predicted values. See Chapter 2 or the glossary at the end of the report for full names and descriptions of the projects.

[a]Model 1: mean expenditure per hectare and project category are expressed as separate variables. $F(13, 42) = 6.64$ (p>.0000); $R^2 = 0.38$.

[b]Model 2: mean expenditure per hectare is expressed separately for each project category. $F(12,43) = 3.45$, (p> .0000); $R^2 = 0.38$.

*Statistically significant at 10 percent.

**Statistically significant at 5 percent.

***Statistically significant at 1 percent.

visible signs of rills and gullies. This is a rough measure that cannot identify imperceptible sheet erosion processes, but it is sufficient to identify any form of rill or gully erosion. Inhabitants of the study villages accompanied the field investigators to tell them the tenure status of the land (private, Forest Department, or government revenue). The transect scoring system is the same as in the drainage line transect, with 1 = low erosion, 2 = medium, and 3 = high erosion. The score was recorded for each segment along with the length of the segment. Weighted averages of the erosion score for each segment can then be summed to give aggregate village-level scores, which can be expressed either as an overall score for the entire village or as separate scores for different land uses and different land capability classifications.

In Table 6.4, the mean transect scores for uncultivated land show no significant differences across project categories, nor are any other village characteristics significantly associated with the erosion score in the bivariate tabular analysis. In fact the scores are marginally better in nonproject villages than they are for each of the other categories. One possible explanation for this is that nonproject villages are generally on flatter land than project villages and less susceptible to erosion. Such a finding might also turn up if the project villages were intentionally selected because they had the most problems: in that case nonproject villages might be in the best condition today because they were in the best condition when the projects began.

Econometric analysis is required to gain a more detailed understanding of the determinants of soil erosion on uncultivated land. The unit of observation for the econometric analysis is the transect segment rather than the village-level average, controlling for land tenure status. The model specification accounts for the fact that observations "clustered" within villages are not independent of each other (Stata Corporation 1999). An ordinal probit model is used because the transect scores are ordinal values of 1, 2, or 3. The observations are not weighted by segment length because that would prohibit weighting them for sampling weights and clustering.[25]

The variables used in the econometric analysis are nearly the same as those used for the determinants of the drainage line scores. One additional variable that was not applicable in the drainage-line analysis is the ownership status of the segment of land in question; a dummy variable indicates whether land is private or common (government revenue or Forest Department land).

The regression results are presented in Table 6.5. As in the drainage line analysis,

Table 6.4 Erosion scores for uncultivated land from the village transect, by project category

Project category	Erosion score
All villages	2.29
NWDPRA	2.41
Jal Sandharan	2.32
NGO2.25	
AGY or IGWDP	2.37
No project	2.15

Source: Authors' 1997 survey data.

Notes: Possible transect scores are 1, 2, and 3. Strictly speaking, they are ordinal variables, not cardinal, but the average scores are shown here for ease of presentation. An average score of 2.00 means average condition is intermediate; less than 2.00 means erosion is low on average, and greater than 2.00 means erosion is high on average.

Kruskal-Wallis ordinal variables test shows no significant difference between project categories or any other village characteristics. See Chapter 2 or the glossary at the end of the report for full names and descriptions of the projects.

[25]When the analysis was conducted without accounting for the survey data, there was practically no difference when weighted or unweighted by segment length.

Table 6.5 Transect scores for the erosion status of uncultivated lands (ordered probit regression)

Variable	Coefficients in model 1[a]	Coefficients in model 2[b]
Altitude range (thousands of meters)	0.230	0.330
	(1.440)	(1.400)
Distance to nearest bus stop in 1987 (kilometers)	0.000	-0.020
	(0.050)	(0.050)
Paved road in 1987 (dummy)	0.360	0.310
	(0.330)	(0.330)
Population density in 1990 (hundreds of persons/square kilometer)	-0.650	-0.660
	(0.220)***	(0.210)***
Distance to *taluka* headquarters (tens of kilometers)	0.050	0.040
	(0.100)	(0.090)
% of inhabitants working in nonagricultural sector	0.005	0.008
	(0.016)	(0.017)
% of inhabitants working primarily as shepherds	0.060	0.040
	(0.050)	(0.050)
Whether the land is operated privately (dummy)	-0.590	-0.570
	(0.360)	(0.360)
Mean project expenditure per hectare (thousands of Rs)	-0.170	...
	(0.060)***	
Project dummy variables for:		
NWDPRA	-0.410	...
	(0.780)	
DPAP and Jal Sandharan	-0.340	...
	(0.460)	
NGO	-1.910	...
	(0.750)**	
AGY or IGWDP	-1.180	...
	(0.720)	
Mean village expenditure per hectare (thousands of Rs)		
Project operated by:		
NWDPRA	...	-0.200
		(0.140)
DPAP and Jal Sandharan	...	-0.200
		(0.070)***
NGO	...	-0.350
		(0.170)**
AGY or IGWDP	...	-0.450
		(0.130)***

Source: Authors' 1997 survey data.

Notes: Observations: 174 from 70 villages. Possible transect scores are 1, 2, and 3. Coefficients and standard errors are adjusted to account for sampling weights and stratification. Predicted values based on the multinomial logit regression in Table 5.2 are used for the four project category variables. Standard errors are not adjusted for use of predicted values. See Chapter 2 or the glossary at the end of the report for full names and descriptions of the projects.

[a]Model 1: mean expenditure per hectare and project category are expressed as separate variables. $F = (13, 42) = 3.45, p > .001$.

[b]Model 2: mean expenditure per hectare is expressed separately for each project category. $F(13, 42) = 3.45, p > .002$.

*Statistically significant at 10 percent.

**Statistically significant at 5 percent.

***Statistically significant at 1 percent.

the econometric analysis is conducted twice. First, the project category variables are specified separately from the funds invested per hectare and then they are combined and allowed to interact with each other. When they are specified separately, all the project category variables have a negative coefficient, indicating that the extent of erosion is reduced. The coefficient is significant only for the NGO category, but it is nearly significant for the AGY or IGWDP category.[26] This is consistent with the finding that these projects are more successful than others in protecting the common lands, and also that many of the villages where NGOs operate have no common land. Expenditure per hectare is statistically significant, suggesting that the specific project category may be less important than the fact that at least some kind of investment takes place. The only other significant variable is population density; higher density indicates lower erosion, which is consistent with the induced innovation hypothesis that land will be better managed when it is scarce. Infrastructure variables are insignificant. Neither the property rights status of the plot nor the number of shepherds in the village is significant, but both are nearly so and both have the expected sign.

The results are similar when the predicted project category and expenditure per hectare are allowed to interact. In this case expenditure under any project reduces erosion, but it is only statistically significant for the NGOs, the AGY or IGWDP, and the DPAP. It is nearly significant for the NWDPRA. The NGO and AGY or IGWDP coefficients have a much greater magnitude than those of the other projects.

The use of actual project dummy variables (not shown) yields similar results. All project categories have statistically significant coefficients, but the degree of significance and the magnitude of the coefficient

Table 6.6 Village-level change in availability of various products from common revenue lands between 1987 and 1997, by project category

Project category	Number of villages	Villages with diffeent directions of change in availability of grass fodder (%)			Villages with different directions of change in availability of tree fodder (%)			Villages with different directions of change in availability of fuel (%)		
		More	Same	Less	More	Same	Less	More	Same	Less
All villages	40	20	28	53	8	63	33	5	60	35
NWDPRA	6	17	17	67	17	50	33	10	60	30
Jal Sandharan	10	30	10	60	20	60	20	0	50	50
NGO	4	25	50	25	0	50	50	13	50	38
AGY or IGWDP	8	25	38	38	0	63	38	0	75	25
No project	12	8	33	58	0	75	25	5	60	35

Source: Authors' 1997 survey data.

Notes: This analysis covers only those villages in the sample that had common revenue land in both 1987 and 1997. The number of villages in each project category is as follows: NWDPRA, 6/10; DPAP, 10/17; NGO, 4/12; AGY or IGWDP, 8/14; no project, 12/17. Kruskal-Wallis test for ordinal variables was conducted to identify variables significantly associated with changes in fuel and fodder supply. The tests show that fuel and fodder supply do not vary significantly across project categories or any other village-level characteristics. Numbers may not add to 100 because of rounding. See Chapter 2 or the glossary at the end of the report for full names and descriptions of the projects.

[26]The standard errors are not corrected for the use of predicted project categories.

remain higher for NGOs and the AGY or IGWDP category.

Change in Availability of Fuel and Fodder from the Common (Government Revenue) Lands

Information on products collected from the common lands was obtained as part of the village-level survey. Respondents were asked in groups about what kinds of products were available in 1997, what kinds were available in 1987, whether the quantity had changed between 1987 and 1997, and if so did it increase or decline. The questions covered grass fodder, tree fodder, fuel, timber, and building materials, and respondents mentioned several other products. However, only grass fodder, tree fodder, and fuel were found in more than a few villages, so the analysis presented here is restricted to those commodities. The mean values of responses by project category, presented in Table 6.6, cover only the subset of 40 Maharashtra villages that had government revenue land in both 1987 and 1997.

No significant differences were found across project categories or any other village characteristics. For grass fodder, most villages reported that there was less available in 1997 than in 1987, and this was the case for all project categories except NGOs, which had closer to the same amount in both years. For tree fodder and fuel, most villages reported having the same amount in both years, but more villages reported a decline than an increase. It appears that the watershed projects have had difficulty in raising availability of these products on the government revenue lands.

Econometric analysis is needed for a more thorough examination of the determinants of changes in access to products of the commons. Ordered probit models are used to analyze the determinants of whether a village has more, less, or the same amount of grass fodder available from the government revenue lands. The explanatory variables are the same as in the analysis of the condition of the drainage line, with the addition of a dummy variable indicating whether grass fodder was available in 1987. As in the earlier analyses, the model is run both with the predicted project category variables expressed separately from the expenditure per hectare, and with them together.

The projects seem to have led to reduced access to grass fodder, compared with nonproject villages (Table 6.7). The variables for expenditure per hectare and the AGY or IGWDP and Jal Sandharan and DPAP project categories have negative, statistically significant signs. Where the expenditure and project category variables interact, the AGY or IGWDP and DPAP variables remain significantly negative, while the other project categories are insignificant.[27] The AGY or IGWDP coefficient also has a much higher magnitude.

This finding is consistent with the earlier findings showing that the AGY and IGWDP were particularly successful in restricting access to common lands and reducing erosion in the drainage line and pasture lands. Improving the condition of these lands requires that access to them be restricted, and the results suggest that access was in fact still restricted at the time of the survey (Table 6.7). Several other variables are significant as well. Population density has a negative sign, while the variables with positive signs include availability of grass fodder in 1987, altitude range, distance to the nearest bus stop in 1987, percentage of households working primarily outside of agriculture, and percentage of households working primarily as shepherds. The highly significant, strongly positive coefficient for shepherds is again

[27]The standard errors are not corrected for the use of predicted project categories.

Table 6.7 Econometric analysis of determinants of change in availability of grass fodder and fuel on government revenue lands (ordered probit regression)

	Gross fodder		Fuel	
Variable	Model 1[a]	Model 2[b]	Model 1[a]	Model 2[b]
Availability of grass fodder (fuel) in 1987	2.13	2.09	1.92	1.46
	(0.64)***	(0.61)***	(0.55)***	(0.83)*
Altitude range (hundreds of meters)	3.76	3.71	0.04	0.02
	(1.26)***	(0.96)***	(0.01)***	(0.11)**
Distance to nearest bus stop in 1987 (kilometers)	0.34	0.53	-0.10	-0.08
	(0.16)	(0.19)***	(0.17)	(0.15)
Paved road in 1987 (dummy)	0.75	0.92	-0.40	-0.36
	(0.56)	(0.66)	(0.51)	(0.42)
Population density in 1990 (hundreds of	-0.75	-1.06	-0.64	-0.36
persons/square kilometer)	(0.23)***	(0.51)**	(0.41)	(0.21)*
Distance to *taluka* headquarters (tens of kilometers)	0.05	0.33	-0.06	-0.06
	(0.04)	(0.33)	(0.04)*	(0.02)*
Inhabitants working primarily in nonagricultural	0.11	0.10	0.04	0.04
sector (%)	(0.03)***	(0.04)**	(0.03)	(0.02)**
Inhabitants working primarily as shepherds (%)	0.69	0.62	0.13	0.12
	(0.19)***	(0.17)***	(0.18)	(0.13)
Mean project expenditure per hectare (thousands of Rs)	-0.43	...	-0.71	...
	(0.17)**		(0.37)*	
Project dummy variables for:				
NWDPRA	0.90	...	7.56	...
	(2.70)		(2.80)**	
DPAP and Jal Sandharan	-2.40	...	5.15	...
	(1.41)*		(1.6)***	
NGO	7.95	...	7.75	...
	(12.75)		(6.62)	
AGY or IGWDP	-5.23	...	2.26	...
	(2.06)**		(2.41)	
Mean village expenditure per hectare (thousands of Rs)				
Project operated by:				
NWDPRA	...	0.06	...	0.60
		(0.60)		(0.36)
DPAP and Jal Sandharan	...	-0.89	...	0.32
		(0.31)***		(0.16)**
NGO	...	1.35	...	-0.71
		(2.29)		(1.39)
AGY or IGWDP	...	-2.04	...	-0.33
		(0.38)***		(0.31)

Source: Authors' 1997 survey data.

Notes: Observations: 40. Possible transect scores range from 1 to 3. Coefficients and standard errors are adjusted to account for sampling weights, stratification, and finite population size. Predicted values based on the multinomial logit regression in Table 5.2 are used for the four project category variables. Standard errors are not adjusted for use of predicted values. See Chapter 2 or the glossary at the end of the report for full names and descriptions of the projects.

[a]Model 1: mean expenditure per hectare and project category are expressed as separate variables.

[b]Model 2: mean expenditure per hectare is expressed separately for each project category. *F*-statistics: fodder: $F(13,19) = 6.27$ (Model 1); $F(13,19) = 6.88$ (Model 2); fuel: $F(13,19) = 2.94$, $p > .02$ (Model 1); $F(13,19) = 2.06$, $p > .08$. (Model 2).

*Statistically significant at 10 percent.

**Statistically significant at 5 percent.

***Statistically significant at 1 percent.

consistent with the finding that it was more difficult to reduce erosion in the villages with the most shepherds, presumably because access restrictions were difficult to enforce. The positive sign for altitude range may reflect high rainfall, which is omitted because it is highly correlated with altitude range. High rainfall stimulates rapid growth of natural vegetation, so it may be that access restrictions can be less strict in these villages. The negative sign for population density either means that availability of fodder has declined as a result of population pressure, or that densely populated villages were more likely to impose access restrictions. The positive sign for the percentage of households working outside of agriculture means either that this caused less competition for fodder, or that there was less pressure to impose restrictions in these villages.

When actual project dummy variables are used (not shown), the result is very similar when the project dummies and expenditure per hectare interact. When they are not allowed to interact, all of the project variables have positive signs but none are significant (while expenditure per hectare is negative and statistically significant).

Table 6.7 also shows the determinants of changes in availability of fuel from government revenue lands. Most variables have the same signs as in the model for changes in grass fodder, and most of the same variables are significant. The coefficient for distance to the *taluka* headquarters is negative and significant for fuel, but positive and insignificant for fodder. The reason for this difference is not clear. One notable difference is that the NWDPRA and DPAP and Jal Sandharan project categories have significantly greater fuel availability than the nonproject villages, while the AGY or IGWDP villages have less. The finding for AGY or IGWDP is consistent with that for grass fodder, while that for the government projects could signify that they succeeded in planting trees and getting them established, but then did not enforce their protection. It is important to note that these results are not duplicated when actual project dummies are used; in that case the NWDPRA coefficient is insignificant and all the others are negative.

The results for tree fodder are similar to those for fuel, which is not surprising since trees are the main source of fuel. These results are not shown.

The strong finding of reduced availability of fodder from the common lands deserves more detailed investigation. Findings from qualitative investigations provide further insight into this issue. In particular, women and livestock herders in many project villages complained in group interviews that they had suffered from loss of access to common lands sealed off to promote regeneration.

It bears repeating that the temporary closure of common lands is expected to lead to benefits for the people who rely on those lands. The finding that these people are the ones who complain about the watershed projects may be an artifact of timing: the interviews took place while the regeneration process was still under way. On the other hand, it also is worth noting that the regeneration process is a long one, and that a number of people may suffer significantly from loss of access to common lands. Project employment will help overcome the associated difficulties, but employment must be substantial enough to compensate for lost income, and it needs to be sustained for several years.

Project Impacts on Livestock Herders

Livestock herders in many project villages complained that they had suffered from loss of access to their traditional grazing lands, which were sealed off to promote regeneration. All of the projects had provided employment opportunities to the herders, but they said it was not enough to compensate them for their losses. This problem commonly arose in Maharashtra, where landless people (typically scheduled castes and tribes) are a small minority in most villages and the decision to close the common lands was usually based on a majority-rule vote. In the IGWDP villages, the decision to begin the project was based on consensus, but some landless people stressed in the group interviews that it was not feasible for them to stand up to the will of a more powerful majority.

In some villages herders said that they had been promised that access restrictions would be temporary, while vegetation was allowed to regenerate. However, they complained that regeneration had already taken place yet the common lands remained off limits to them. Such inequity is more likely to be a problem where projects succeed in productivity and environmental objectives.

Table 6.8 Percentage of respondents in Maharashtra who say they benefited from the watershed project, by project category and landholding size

		Landholding size category			
Project category	All land sizes[a]	Landless	0-1 hectares	1-2 hectares	> 2 hectares
All projects[b]	26	12	19	26	45
NWDPRA	8	0	17	0	17
DPAP and Jal Sandharan	17	0	0	33	20
NGO	39	29	44	25	63
AGY or IGWDP[c]	31	14	0	33	60

Source: Authors' 1997 survey data.

Notes: Findings based on household survey; 120 respondents from 10 villages. See Chapter 2 or the glossary at the end of the report for full names and descriptions of the projects.

[a]For all project categories together, perception of benefits differs significantly across landholding categories (chi-square 8.82, 3 df, $p < .03$).

[b]For all landholding sizes together, perception of benefits differs significantly across project categories (chi-square 8.51, 3 df, $p < .036$).

[c]Within project categories, perception of benefits differs significantly across landholding sizes only for the AGY or IGWDP projects (chi = square 8.08, 3 df, $p < .044$).

Table 6.9 Percentage of respondents in Maharashtra who say they were harmed by the watershed project, by project category and landholding size

		Landholding size category			
Project category	All respondents	Landless	0-1 hectares	1-2 hectares	> 2 hectares
All projects	11	19	8	10	7
NWDPRA	4	0	17	0	0
DPAP and Jal Sandharan	13	33	0	11	0
NGO	8	14	0	8	13
AGY or IGWDP	17	29	14	17	10

Source: Authors' 1997 survey data.

Notes: Findings based on household survey; 120 respondents in Maharashtra. Perceptions of harm by project categories are statistically insignificant across project and landholding categories. See Chapter 2 or the glossary at the end of the report for full names and descriptions of the projects.

In other places, herders were able to ignore grazing restrictions, protecting their immediate livelihoods but undermining project objectives. These findings from qualitative discussions are consistent with the result in the quantitative analysis that a high population of shepherds raised the extent of erosion but also raised access to grass fodder, compared with other villages. To reiterate, this does not necessarily mean that these villages are more productive, just that grass fodder from the commons was more readily available at the time of the survey, possibly due to lack of restrictions.

Additional data from open-ended questions at the household level support these findings. In 13 Maharashtra villages, respondents' perception that they had benefited from the projects generally rose with landholding size (Table 6.8). Landless people were much more likely to indicate that the project had harmed their interests; among landless people the unanimous complaint was lost access to common lands (Table 6.9).

As revealed in Chapter 5, some NGOs and the Jal Sandharan project in Maharashtra appear to have dealt with this problem by selecting many villages that had no government revenue land, thus avoiding the issue. While this approach may be sensible as a way to avoid equity tradeoffs, obviously it provides no lessons about how to address the problem in the majority of villages that do have government revenue land.

On the other hand, a few NGOs, such as Chaitanya and MYRADA in Andhra Pradesh, have explicitly aimed to develop innovative solutions to the problem of managing common lands. They try to build the interests of different groups into the project design at the outset. For example, in some projects landless people are granted fishing rights in the water bodies protected by soil conservation and revegetation of the common lands. Unlike in Maharashtra, landless and near-landless respondents in Andhra Pradesh unanimously reported having benefited from NGO projects.

Social Centre, a Maharashtra NGO, grants fishing rights to landless people in some villages, including Mendhwan, which is covered under the current study (WOTR 1999). Some projects encourage farmers without irrigation to dig group-owned wells so that they have an interest in promoting groundwater recharge. Outside of the study area in the famous Sukhomajri and *Pani Panchayat* projects, landless people even own rights to water for tank or lift irrigation, which they utilize by leasing in farmland or, in the case of Sukhomajri, trade to other farmers (Chopra, Kadekodi, and Murty 1990; Patel-Weynand 1997). And in several Andhra Pradesh villages not covered by any kind of project, shepherds lease cultivated land and manage it as pasture. Such an arrangement could be made in a watershed project: if shepherds had exclusive rights to grazing lands they would have an incentive to invest in raising their productivity, and this would likely include reduced grazing pressure and thus reduced erosion. A wide assortment of such arrangements can be devised to spread the

Table 6.10 Reported changes in number of days of employment between 1987 and 1997, by project category

	% of respondents indicating more, less, or same access to employment		
Project category	More	Same	Less
All villages	33	61	6
NWDPRA	9	91	0
Jal Sandharan	29	65	6
NGO	43	47	10
AGY or IGWDP	72	17	11
No project	18	78	4

Source: Authors' 1997 survey data.

Notes: Findings from household-level interviews: n = 85. Thirty-five respondents who do not engage in wage labor did not respond. Kruskal-Wallis test for ordinal variables shows that change in number of employment days varies significantly across project categories (chi-square = 13.6, 4 df, $p < .01$). See Chapter 2 or the glossary at the end of the report for full names and descriptions of the projects.

benefits of watershed development and, as a consequence, increase its chances of success.

Watershed agencies argue that if their work is successful, landless people will benefit in the long term. In the famous Adgaon watershed, annual employment rose from 75 to 200 days, and laborers' incomes rose above those of small farmers according to an NGO involved in the project (WOTR 1999). Social Centre found that after four years of watershed management, laborers in Mendhwan Village could find employment eight months of the year, compared with three months work previously. In Sherikoldara, landowners began to lease land to laborers rather than pay the high wage costs (WOTR 1999).

Respondents in this study were asked whether they obtained more, less, or the same number of employment days than before the project period. No distinction was made between short-term work generated as part of the project and long-term changes in demand for labor. Respondents in the AGY, IGWDP, and NGO project villages indicated with much greater frequency that employment opportunities had risen compared with other villages (Table 6.10).

Project Impacts on Women

During unstructured interviews in 35 of the study villages, women brought up a number of problems they had encountered as a result of the projects. Reduced access to firewood was the most frequently expressed concern. Women in half the villages indicated that they had to walk farther than before to collect fuel. Many said that they were more likely to use alternative sources, such as purchasing fuel from the market or even stealing it from land belonging to other villagers.

Women from one-third of the villages surveyed also complained that closing the commons deprived them of several sources of income that they controlled independently. When they had access to the commons, they said, they sold brooms made out of grasses that grow in the common lands and collected marketable products such as tamarind pods or medicinal plants. Many poorer women keep a small number of goats; when the commons were open, these animals could be kept virtually free of cost. A woman who went to the commons to collect grass for brooms, for example, could let her goat roam while doing so. When the commons were closed, however, goats could no longer graze and the women had to collect food for them. Indirectly this reduced the income a woman could earn on each trip into the common lands. Women either had to work harder to keep the goats or give them up.

Women reported similar effects under all projects that succeeded in closing common lands. This means that the problem was most severe in the NGO/government collaborative projects. None of the projects made any special effort to replace women's lost sources of income, although some tried to train women in other activities, such as the use of improved cook stoves, tailoring, or raising plants and trees that could be used in the watershed program. Training activities were twice as likely to be found in the projects with an NGO component than in the purely government projects.

Watershed development can also have negative impacts on women from wealthy households owning irrigated lands. Although increased crop production from expanded irrigation provides benefits, it may require that women contribute more agricultural labor. On top of household work and child rearing duties, this could be a significant problem. Although some women mentioned it in open-ended discussions, detailed data are not available.

Some of the negative effects on women could be overcome if a greater effort was made to include them in decisionmaking. The village-level watershed committees that helped plan and organize the watershed work were largely made up of men. In the government projects all the committee members were men, while projects with an

NGO component typically had one or two women out of a total of 7–10 members. Usually these women did not assert themselves in the meetings. For women committee members to have much influence will probably require giving them half the positions on the committee. In addition, special efforts will be needed to schedule meetings around women's domestic duties and to organize childcare.

All of the project categories ignored women when it came to training villagers to participate in watershed development, at least until the time came to explain to them that they would lose access to common lands. Typically, only men received training in the technical work related to watershed development. Without technical training, women were not eligible for the better paying jobs such as helping lay contour lines or counting the number of loads of earth moved, even though they were just as qualified as men.

To make watershed development projects more hospitable to women, not only should more women staff members be hired, but male staff members should be trained to be sensitive to gender concerns. Since male staff members interact with men in the village, they are the logical ones to try to make the men more receptive to a role for women in decisionmaking. Women need to be included in discussions from the very beginning, and they need a voice in determining whether to accept a project. In most programs this decision is taken on the basis of a vote, with consent from two-thirds of households required to proceed. However, the male head of household is the one who votes: it is assumed that women in the household will share his interests. Clearly, women need an independent vote.

CHAPTER 7

Promoting Irrigation Development

Raising the water table to promote irrigation development is a primary objective of most watershed projects operating in Maharashtra, and a secondary objective of those in Andhra Pradesh. The projects seek to raise the water table through soil and water conservation and revegetation measures, which encourage rainwater to infiltrate into the soil, gradually augmenting groundwater. Primary among the soil and water conservation investments are structures to impound water, placed in the main drainage lines. They range from small "gully checks" to major structures such as percolation tanks or check dams. Outside of the drainage line, contour trenches are dug along uncultivated hillsides.

This chapter examines the impact of the projects' efforts to promote irrigation development. Data limitations concerning the irrigation potential of each village make it impossible to estimate precisely the contribution of each watershed project, so the analysis relies on more than one source of information. These include changes in the number of seasons over which crops are irrigated on the sample plots (cropping intensity) and farmers' perceptions of whether water-harvesting measures raised the water table.

It is important to acknowledge weaknesses in the data used in this analysis: not all of the many factors that determine irrigated area could be incorporated. In particular, the hard-rock aquifers of the Deccan Plateau are known for high spatial variation in irrigation potential. In some villages—or in some areas within some villages—the potential for raising irrigated area is quite favorable, but in others it is minimal. Unfortunately no data are available on the nature of aquifers in each village.

Changes in Net Irrigated Area

At the village level, the relevant measure of increased irrigation is the change between 1987 and 1997 in the percentage of cultivated area that is irrigated.[28] Official records contain information only on irrigation infrastructure, not irrigated area per se, so it provides no information on seasonal variation in availability of groundwater for pumping. In addition, the data actually obtained from official sources for this study contained major inconsistencies and so could not be used for analysis.

Plot-level data provide more useful information about irrigation development. In particular, data on changes in cropping intensity give additional details regarding more subtle changes

[28]Virtually all arable land was under cultivation by 1987; the total area under cultivation was roughly constant between 1987 and 1997

in irrigated area. For example, a plot that was irrigated for one season in 1987 may be irrigated for two seasons in 1997, but the village level data on irrigation infrastructure would not show the change.

The indicator for increased cropping intensity measures the change in the number of seasons when irrigation was used for each plot in the sample. For example, if a plot was rainfed in 1987 but irrigated for two seasons in 1997, its score is +2. This information is collected through recall of the plot's owner.

Irrigation intensity increased much more in Maharashtra than Andhra Pradesh, with a mean increase of 0.35 in Maharashtra, compared with 0.20 in Andhra Pradesh (Table 7.1). The difference across project categories is significant only in Andhra Pradesh, where plots under the World Bank project had the highest increase in cropping intensity. In Maharashtra, plots under the AGY and IGWDP had the highest irrigation increase, but this difference was not statistically significant. In Andhra Pradesh, nonproject villages had a higher mean increase in irrigation intensity than all projects except the World Bank or ICAR projects.

Table 7.1 Mean increase in number of seasons plots were irrigated, 1987–97, by project category

	Mean increase in seasons irrigated[a]	
Project Category	Maharashtra	Andhra Pradesh
All plots	0.35	0.20
NWDPRA	0.37	0.10
DPAP or Jal Sandharan	0.33	0.00
NGO	0.25	0.11
AGY or IGWDP	0.44	n.a.
World Bank or ICAR	n.a.	0.49
No project	0.25	0.21

Source: Authors' 1997 survey data.

Notes: See Chapter 2 or the glossary at the end of the report for full names and descriptions of the projects. N.a. indicates not available.

[a]Kruskal-Wallis test shows that differences across project categories are statistically significant, in Andhra Pradesh but not in Maharashtra.

It was hoped that more detailed information about the determinants of increases in irrigation development could be gained from multivariate analysis. Unfortunately, such analysis failed to reveal any additional information—almost certainly due to the lack of data on such important confounding variables as the nature of the aquifer. Regression findings are not presented here because they provided no additional insights.

Respondents' Perceptions of Projects' Effects on Irrigation Development

Qualitative discussions revealed that respondents are keenly aware that water-harvesting structures in the drainage line can raise the groundwater level, thus promoting irrigation development. In several villages they indicated that water levels in open wells had risen visibly following the construction of water-harvesting structures. In some of the Maharashtra villages, however, respondents indicated that low rainfall in recent years made it difficult to discern the effectiveness of water harvesting. And in some villages, respondents reported that certain water-harvesting structures leaked water, making them ineffective.

All the Maharashtra projects focused on water harvesting, whereas in Andhra Pradesh, the World Bank, ICAR, and NWDPRA devoted only minimal attention to it. Only the DPAP focused primarily on water harvesting, and two of the three NGO projects also included water harvesting as a major project activity. In Andhra Pradesh, owners of irrigated land had a good impression of the DPAP's efforts in this regard.

Discussions in both states revealed a sharp sense among farmers of the differences in the structures that could promote water harvesting. For example, the DPAP and Jal Sandharan, for which water harvesting was the main project objective, had large budgets for gully structures, and they built the largest and most solid impermeable structures. The NWDPRA and World Bank projects, on the other hand, were not designed with water harvesting in mind, and so they budgeted much smaller amounts mainly for vegetative or loose stone structures. Respondents were keenly aware of these differences, especially in Maharashtra, where they could compare the NWDPRA gully structures with those built under COWDEP in the 1980s. They did not perceive that the NWDPRA's work had much impact.

A similar issue arose among NGOs. As discussed in Chapter 5, projects vary in the number of technically trained people on their staff. Some NGOs, like Chaitanya, employed no technically trained staff and focused exclusively on social organization, relying on indigenous technical knowledge in the design of their watershed interventions. Some other NGOs, like MYRADA, employed engineers to oversee the technical work. Similar differences were found in Maharashtra. Not surprisingly, respondents reported greater impact on water harvesting where projects employed technical experts. In the Chaitanya village, for example, the water-harvesting structure was not effective because it leaked. Such a finding underscores the philosophy behind the AGY and IGWDP, which sought to combine the technical expertise of government agencies with the social organization skills of NGOs.

In a semi-structured interview that was part of the household survey, respondents were asked to list the kinds of benefits they perceived from the project operating in their village. Table 7.2 shows the number and percentage of respondents who mentioned irrigation benefits. The table records the numbers not only for farmers with irrigation but for all farmers and for all respondents, including those without land. In fact, all of those who reported that they had benefited from the projects were farmers with irrigation; in other words, the projects were not able to help others gain access to water.

Table 7.2 Number and percentage of respondents reporting that water-harvesting investments improved their access to irrigation

Project category	Number	All respondents (%)[a]	Farmers (%)	Farmers with irrigation (%)
Maharashtra	21	18	23	46
NWDPRA	2	8	11	13
DPAP	3	13	17	50
NGO	6	17	21	60
AGY or IGWDP	10	28	37	71
Andhra Pradesh	9	6	8	13
NWDPRA	2	6	8	22
DPAP	3	8	11	18
NGO	2	6	7	9
World Bank or ICAR	2	4	5	11

Source: Authors' 1997 survey data.

Notes: Kruskal-Wallis test shows that differences across project categories are statistically insignificant. See Chapter 2 or the glossary at the end of the report for full names and descriptions of the projects.

[a]This represents the percentage of all respondents who reported better access to irrigation. For example, 21 respondents in Maharashtra represent 18 percent of all respondents.

Farmers with irrigation are best suited to explain whether they thought project activities had helped raise the water table. Figures in the table show that a much higher percentage of respondents perceived benefits in Maharashtra than Andhra Pradesh, and this is consistent with project objectives in the two states. In Maharashtra, irrigated farmers reported the least benefits from NWDPRA, and this is consistent with that project's lack of focus on water harvesting; reported benefits were highest for the AGY and IGWDP projects. In Andhra Pradesh, perceived irrigation benefits were very low for all projects.

One obvious point in the table is that perceived benefits from irrigation are highly concentrated among farmers with access to irrigated land. There are also indirect benefits, such as higher employment demand, which respondents did not mention. In any case, the skewed distribution of the most valuable project benefits to those who already have the most prized asset (irrigated land) is a source of concern to many project officials and other commentators. There has been much discussion of what can be done to distribute project benefits more evenly. For example, in some projects outside the current study area all village inhabitants share equally in water resources generated by the project (see Chapter 6). No project in this study undertook such ambitious steps, but some of them did try to help spread the benefits of irrigation. In particular, the IGWDP agreed to take up work only in villages that agreed not to drill any borewells, which draw more water than traditional open wells and would appropriate harvested water disproportionately. For similar reasons, the IGWDP also insists that no farmers may take up water-intensive crops such as sugarcane in response to higher water supplies. Sugarcane farmers would draw more water from their wells, reducing the water level in other wells. Also, farmers with excess water might choose to sell it to their neighbors if they cannot grow water-intensive crops. A few other NGOs mentioned similar restrictions, but most did not. None of the government projects tried to impose any such restrictions.

The new, common guidelines under the Ministry of Rural Development and the Ministry of Agriculture and Cooperation discuss the need to develop mechanisms for landless and other poor people to share usufruct rights to resources generated under the projects (India, Ministry of Agriculture and Cooperation 2000). However, they offer no suggestions on how to proceed. This is clearly an area where projects would benefit by experimenting to identify useful approaches.

Another approach to sharing the benefits of water harvesting is to help resource-poor farmers invest in their own wells. India has quite a bit of experience in this regard; a centrally sponsored program, for example, digs individual private wells for marginal farmers from scheduled castes and tribes and backward classes. Some projects have invested in group-owned wells, but the most common experience was that groups had difficulty in working together to manage and maintain their wells. This matches the experience of state-owned cooperative tubewells (Shah 1993). In the villages in this study, there were numerous cases of group-owned wells, but in nearly every case they were jointly owned by brothers who inherited the well from their father. There was one recorded case of some neighbors (not relatives) who jointly invested their own funds in a well, but within a few years a dispute emerged and the case ended up in court. Against this backdrop, most projects are hesitant to invest in jointly owned wells. In the current study, only one NGO, Gramayan in Maharashtra, invested funds in a group-owned well. According to respondents it was managed effectively.

CHAPTER 8

Natural Resource Management and Productivity of Rainfed Agricultural Land

Raising the productivity of rainfed agriculture is the most important objective for some watershed projects, particularly the NWDPRA, the ICAR model watersheds, and the World Bank's Pilot Project and IWDP (Plains). It is particularly important where opportunities for water harvesting are limited, as in many Andhra Pradesh project locations. The watershed approach to raising the productivity of rainfed agriculture begins with conserving soil on rainfed plots, which implies efforts to retain soil nutrients and concentrate moisture. This in turn creates opportunities for planting high-yielding varieties that require more water and nutrients. Or in areas with black soils and high rainfall, such efforts may enable farmers to harvest an additional crop each year.

Following a review of projects' policies for subsidizing the development of private land, this chapter investigates the nature and extent of interaction between project staff and farmers, since technical assistance for rainfed agriculture presumably involves working closely with farmers. It then focuses on efforts to conserve soil and moisture, both through improved agronomic methods and investment in soil conservation structures. Analysis of rainfed farmers' adoption of new varieties and their net returns to cultivation follows. Because of the emphasis on rainfed agriculture, the quantitative analysis focuses on plots that were unirrigated in 1987 and 1997.

Project Subsidies to Participants

Subsidies are a contentious and increasingly complex issue in watershed projects. Approaches have evolved over time, with significant trial and error. In the early days of the Bombay Land Improvement Scheme, bunds were installed on some farmers' fields without their consent, yet they were expected to repay the bank for the cost of the work undertaken. They were listed as defaulters if they refused (World Bank 1988).

Most contemporary watershed projects take the opposite approach, with watershed works being heavily subsidized and no thought given to cost recovery. The rationale for this approach was that farmers who benefited from canal irrigation did not have to pay for the canal, so why should farmers of rainfed land, who benefit much less from watershed projects, have to pay for the works undertaken?[29] This argument is really a matter of opinion, but it matters

[29]Recently there has been renewed discussion of the possibility of covering maintenance costs by charging for irrigation.

because these farmers often are not interested in the measures introduced under watershed projects and have no intention of maintaining them once the project ends (Kerr, Sanghi, and Sriamappa 1996; Sanders et al. 1999). Under these circumstances, it is important to require some kind of payment or other sacrifice by "beneficiaries" simply to make sure that they really want the work and are likely to maintain the assets created. Otherwise the project will simply be a waste of money. (This problem does not arise in irrigation projects, because there has never been a farmer in India who did not want irrigation!)

Why Farmers Accept Measures They Do Not Want

There are two main reasons why farmers would allow measures to be taken on their land that they do not really want. The first is that some projects install structures on farmers' plots without their consent, although this practice is diminishing. Watershed officials increasingly appreciate the fact that a structure built in one location can generate on-site costs but only downstream benefits, so in current projects measures are rarely undertaken in farmers' fields without their consent.

Another problem that may lead farmers to accept measures they do not want results from the fact that in most projects, watershed works are labor-intensive and very highly subsidized. All projects covered in this study subsidized work on common lands at the rate of 100 percent, generating ample employment for workers to plant vegetation, dig trenches, and build structures. Even on private land, the typical subsidy rate was 90 percent, and the remaining 10 percent was not paid in cash but in kind (in the form of labor). Moreover, much of the project work was undertaken in the dry season when labor demand was scarce, and in many projects wages exceeded the slack season market wage. So even if a project paid only 90 percent of the subsidized wage, it often represented more than the farmer could earn through other available opportunities. Under these circumstances, it may well make sense for a farmer to accept an unwanted structure on his field, provided of course that the costs of dismantling it are low.

Subsidy Policy and Practice under Each Project

NWDPRA. Project guidelines called for a contribution by farmers only for work undertaken on private lands, except that no single family could receive subsidized assistance worth more than Rs 5,000 (India, Ministry of Agriculture 1991). Specific terms were not mentioned. From the present study, it is not clear how the farmer's contribution worked in practice. In the Maharashtra villages, no work was done on private lands, so the issue did not arise. In Andhra Pradesh, work was undertaken on private lands using labor paid for by the project, but respondents did not indicate that they had contributed anything. Kolavalli (1998) found that the NWDPRA collected a small contribution from farmers in only one of the four project sites he visited.

DPAP and Jal Sandharan. Beneficiaries contributed nothing to these projects. Most work was conducted on nonarable land, but even the minority of work done on private land was entirely subsidized. Usually this work was done using contracted labor. This increased the possibility that the farmer would not be aware that the work was taking place, but it sharply reduced the incentive for the farmer to accept unwanted work. Under the new guidelines the DPAP still offers essentially a 90 percent subsidy for soil conservation work on private land, and because the projects pay the government-mandated minimum wage the actual subsidy is closer to 100 percent.

NGOs. Several NGOs called for a 10 percent farmer's contribution for work on private land, paid in kind (in the form of labor). As mentioned, however, the wage scale was inflated so that employment benefits remained substantial to the farmer. Many NGOs preferred to contract with farmers to do the work on their own fields on the principle that this would raise the quality of the work.

Two NGOs in Andhra Pradesh, on the other hand, required a much more substantial farmer's contribution on private land. Chaitanya required a 50 percent contribution while MYRADA had recently introduced a 33 percent requirement. In some villages not included in this study, MYRADA has experimented with zero subsidies for work on private land (Fernandez 1998). Chaitanya and MYRADA offered lower subsidies in recognition that the farmer would be the primary beneficiary of the work and that farmers would certainly pay attention to its quality if they helped pay for it. There was no contribution for work done on common land.

AGY and IGWDP. In these projects a private landowner's contribution was about 8 percent, but this figure was inflated because dry-season wages under the project often exceeded existing market wages.

ICAR. Under the ICAR model watersheds, all costs were paid by the project and farmers were provided improved seeds and other inputs free of charge. Little or no employment was generated as part of project implementation.

World Bank. The World Bank Pilot Project and IWDP both called for a farmer's contribution of 10 percent on cultivated lands. The contribution was in kind in the form of the farmer's labor. Farmers also received various free inputs such as improved seeds and fertilizers that more than made up for the value of any contribution. There was no cost-sharing for work on common lands.

Interaction Between Project Staff and Survey Respondents

Table 8.1 shows the percentage of farmers from each project category who reported contact with project staff. To distinguish among different types of interaction, it also gives the percentage of respondents who

Table 8.1 Percentage of farmers who interacted with watershed project staff, by project category

Project category	% who interacted	% who received technical recommendations	% who adopted a technical recommendation
Maharashtra			
Overall	30	17	9
NWDPRA	0	n.a.	n.a.
DPAP and Jal Sandharan	0	n.a.	n.a.
NGO	44	25	8
AGY or IGWDP	56	28	19
Andhra Pradesh			
Overall	67	53	50
NWDPRA	56	51	49
DPAP and Jal Sandharan	58	22	22
NGO	70	57	54
World Bank or ICAR	78	72	67

Source: Authors' 1997 survey data.

Notes: This table excludes respondents from nonproject villages. See Chapter 2 or the glossary at the end of the report for full names and descriptions of the projects. N.a. indicates not available.

received technical recommendations related to rainfed agriculture and the percentage who actually adopted a practice recommended by the watershed project staff.

Four main points are worth mentioning. First, overall reported interaction rates were not very high. This reflects the fact that watershed projects rarely cover every farmer's field in every project site. Second, interaction was much higher in Andhra Pradesh than Maharashtra, and it was much more likely to include technical recommendations. This finding reflects the way in which projects operated in the two states. There was less scope for interaction in western Maharashtra because most projects there focused on soil and water conservation on nonarable lands rather than technical interventions on farmers' fields. Also, an important objective of the NGO and NGO/government collaborative projects was to facilitate beneficiaries' access to government services and markets rather than to give direct technical advice.

The picture in Andhra Pradesh is very different. Here, the level of interaction between staff and respondents was much higher, and most of that interaction came in the form of technical recommendations for rainfed agriculture. Only the DPAP, whose primary mission was to develop water resources through groundwater recharge, had low levels of technical interaction. The World Bank and ICAR projects had high levels of interaction, almost all of it in the form of technical recommendations.

A third noticeable finding is that in Maharashtra, all of the interviewed farmers in the NWDPRA and Jal Sandharan project villages said they had never interacted with project staff. This is quite surprising, particularly for the NWDPRA, whose mandate was to promote rainfed agricultural development. Most likely it reflects the focus on water harvesting in the *taluka*-level departments that implemented the project. Even so, at first glance it is surprising that there was no interaction based on employment of labor. But this is explained by the fact that the project officials worked through an intermediary in the village who in turn hired workers, so that there was no explicit interaction between project officials and laborers.

NGOs and the AGY and IGWDP projects in Maharashtra had more interaction with farmers, but the figures are still lower than in Andhra Pradesh. They reflect mainly project efforts to facilitate social organization and to mobilize laborers rather than to provide technical assistance for rainfed agriculture.

A fourth point concerns the percentage of respondents who said they actually had adopted a practice or technology recommended by watershed project staff on the plot in question. These figures quite closely reflect the figures for technical recommendations. In Andhra Pradesh, almost all farmers who received technical recommendations also adopted them. In Maharashtra there was very little adoption of specific technologies.

Adoption of Soil and Water Conservation Practices

Despite the historic focus of most Indian soil and water conservation programs on mechanical measures, soil scientists and agronomists often stress that there is much more to soil conservation than trapping runoff water behind mechanical or vegetative barriers. Conservation begins with sound agronomic practices such as maintaining soil cover and cultivating across the slope to encourage infiltration and reduce runoff. Accordingly, this section examines farmers' adoption of various approaches to conserving soil.

Agronomic Practices

Respondents were asked about a variety of conservation-oriented agronomic practices, including strict contour cultivation, cultivation across the slope, retaining stubble in the plot, and applying mulches to cover bare soil. Of all of these practices, cultivation across the slope was the only one practiced by more than a handful of farmers. Farmers indicated that they recognized the value of applying mulches and retaining stubble in the fields throughout the dry season, but they rarely carried out these practices because of the high opportunity cost of forgoing use of the cut stubble for fuel and feed.

Respondents uniformly said that strict contour farming is impractical except on irrigated land and plots steeper than those covered in this survey. Not a single respondent practiced contour cultivation. This finding echoes the points about contour cultivation made by Kerr and Sanghi (1992). In short, numerous basic features of indigenous rainfed farming systems are integrally linked to quadrilateral plot boundaries, and contour bunds and contour cultivation directly interfere with them. As a result, adopting contour farming carries high opportunity costs. In the 1990s, many watershed projects still officially recommended contour cultivation on rainfed plots, but in practice this was ignored. Efforts were limited to promoting "modified contour cultivation," which simply means cultivating across the slope. Project staff ignored the official instructions to promote contour cultivation because farmers simply would not adopt it, except on irrigated and very steep rainfed plots.

Given the mild slopes and small area of plots in the sample,[30] cultivation across the slope is virtually as effective as strict contour cultivation. Farmers indicated that they traditionally alternated the direction of cultivation each year, going along one boundary one year and the other boundary the next. Where plots were long and narrow, with the long side running along the slope, many farmers cultivated along the slope every year to reduce the number of turns they had to make during field operations. This led to increased runoff and erosion.

Data presented in Table 8.2 suggest that watershed projects were effective in encouraging farmers to cultivate across the slope, particularly in Andhra Pradesh where interaction between respondents and watershed project staff was high. Farmers were significantly more likely to cultivate across the slope in project villages, particularly if they had interacted with project staff. In Maharashtra there was no significant difference between project and nonproject plots.

Regressions to explain adoption of cultivation across the slope did not yield interesting results and so are not presented. Using predicted project dummy variables, the only statistically significant variable is the dummy indicating whether the farmer interacted with the project staff. No plot, household, or village characteristics are statistically significant. With actual project dummy variables, the project categories are statistically significant as long as the dummy variable for interaction with project staff is omitted. When that variable is included, it dominates the effects of the project categories, and none of them are significant.

In any case, the tabular analysis strongly suggests that projects were effective in encouraging farmers to cultivate across the slope. Farmers in villages covered by the World Bank specifically mentioned this as one of the two most important project benefits. (The other was the introduction of new seeds for improved varieties and horticultural crops.) Since numbers were high for respondents who did not interact with project staff as well as for those who did, it seems likely that the message about cultivating across the slope spread from farmer

[30]Of the plots surveyed, 95 percent had less than a 4 percent slope, and the median size of rainfed plots was exactly 1.00 acre (0.42 hectare).

Table 8.2 Percentage of farmers who cultivated across the slope, by project category, both states

	All respondents		Respondents who interacted witht he project staff	
Project category	Total observations	% of total	Total observations	% of total
Maharashtra				
Overall	86	65	24	58
NWDPRA	12	50	0	...
DPAP and Jal Sandharan	10	90	0	...
NGO	24	63	12	58
AGY or IGWDP	21	62	12	58
No project	19	68	...	...
Andhra Pradesh				
Overall	127	80	76	96
NWDPRA	25	72	13	100
DPAP	27	100	19	100
NGO	22	86	17	94
World Bank or ICAR	33	88	27	93
No project	20	40	...	...

Source: Authors' 1997 survey data.

Notes: This table covers only plots with a slope. Kruskal-Wallis test indicates that project category is significantly associated with cultivation across the slope for Andhra Pradesh but not Maharashtra. Maharashtra: chi-square = 0.12, 4 df (highly insignificant); Andhra Pradesh: chi-square = 28.83, 4 df, $p < .001$. See Chapter 2 or the glossary at the end of the report for full names and descriptions of the projects.

to farmer, even beyond the scope of the project.

It is surprising that such simple technical assistance as cultivation across the slope could have such a big impact; it is also ironic that beneficiaries mentioned it as one of the most important benefits of an extremely costly watershed development project. The technology of cultivation across the slope costs nothing monetarily, and it can be readily disseminated without a watershed project.

Investment and Maintenance of Soil and Water Conservation Structures

Data on total soil and water conservation investment expenditure between 1987 and 1997 were collected for each plot covered under the study.[31] It is important to stress that expenditure is not synonymous with protection against erosion, for two reasons. First, plots vary in their susceptibility to erosion because agroclimatic factors (like slope, soil type, and rainfall) vary and because conditions at the start of the study period are different. Therefore, one plot may require more investment than another for protection against erosion. Second, there are many ways to protect against soil erosion, and the effectiveness of the various ways is not necessarily related to the cost. Vegetative barriers are less expensive than earthen barriers, for example, and agronomic practices like cultivation across the slope cost little or nothing. Despite this

[31] This section draws on material previously published in Kerr et al. (1999).

caveat, investment levels do provide useful information about what both projects and farmers are doing to control erosion, and how project interventions affect farmers' own investments. This in turn can help policymakers and watershed officials target their interventions to support the kinds of investments that farmers are less likely to make with their own funds.

The focus here is on soil and water conservation investments on rainfed plots, since the evidence suggests that irrigated plots receive plenty of investment with neither financial nor technical assistance. The types of soil conservation investments listed by respondents include land leveling, grass strips, drains, tree planting, and earthen, stone, or vegetative barriers. The mean value of total investment between 1987 and 1997 on all rainfed plots was about Rs 4,475 per hectare in real terms (Rs 35 = US$1.00 in 1997). The corresponding value for irrigated plots was Rs 69,900, of which Rs 10,630 was for nonirrigation investments like leveling and bunding.

Figure 8.1 makes three main points about the variation in both total investment and source of finance across project categories. First, plots under NGO/government projects and in nonproject villages had the highest levels of investment, followed by those in government projects. Second, while NGO/government and government projects invested about the same amount in subsidies, farmers themselves contributed a much higher amount to the NGO projects. Third, farmers used very little credit to finance their investments, but this amount was much higher in nonproject villages that had little if any access to subsidies for soil and water conservation. (A few farmers in nonproject villages received subsidies from sources other than watershed projects.)

This initial picture of total investment suggests that watershed projects did not succeed in stimulating soil and water conservation investments that farmers would not have made otherwise. However, the situation changes when one looks at how investment costs and sources of finance

Figure 8.1 SWC investment by project category and source of finance

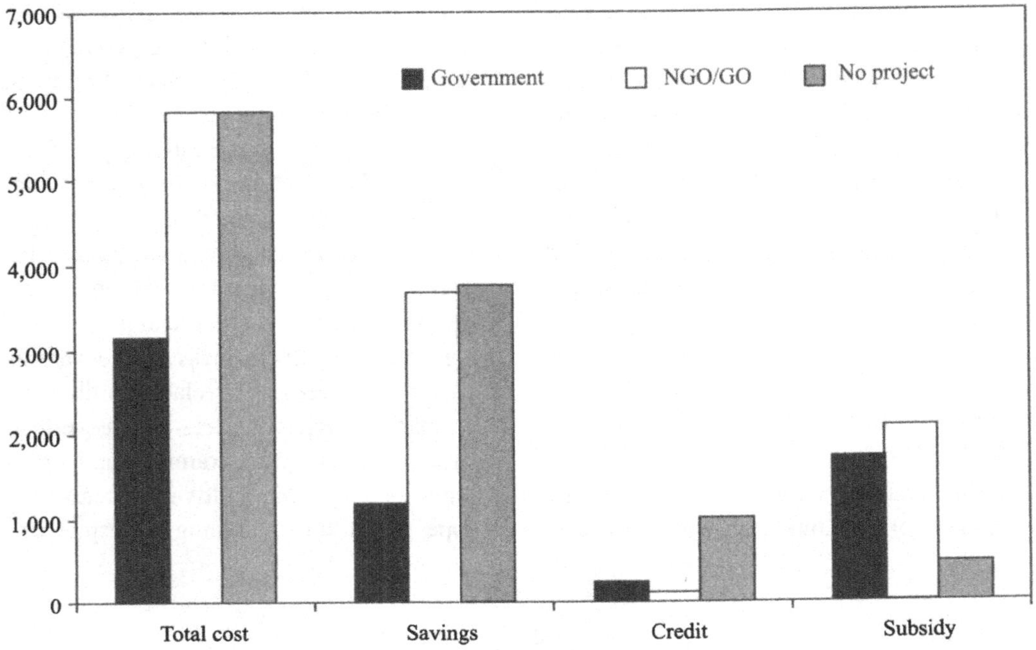

varied by the slope of the plot. While total investment cost varied somewhat by slope, the source of finance differed dramatically (Figure 8.2). Farmers invested their own savings mainly on plots with less than 2 percent slope. They used credit exclusively on these plots (not shown in the figure). Watershed agencies, meanwhile, devoted their funds mainly to plots with more than 2 percent slope. The reason behind this finding is most likely that soil and water conservation investments have important effects on productivity in semi-arid rainfed agriculture, and efforts to conserve and concentrate soil and water may have greater effects on productivity on flatter plots with more fertile soils. As a result, that is where farmers invested their own funds. This clearly suggests that funds from watershed projects complement farmers' own investments by investing on sloped plots that farmers would otherwise neglect.

The question remains why farmers in nonproject villages or those covered by the NGO or NGO/government collaborative projects all invested more than farmers in the villages covered by the government projects. There are at least four reasons to consider. First, it may be that the government projects selected villages where farmers were less able or interested in investing on rainfed plots, but the data provide no obvious indication that this should be so. A second possibility is that vegetative technologies under the NWDPRA and World Bank cost less than those introduced under the NGO and NGO/government projects. But this does not explain the small proportion of total investment costs paid by farmers with their own funds. Third, it may be that farmers in government project villages invested less of their own funds because they were waiting for the government to pay for the investments, which would be reasonable given that government projects generally subsidized their projects 100 percent. Fourth, the higher cost-sharing requirements of some NGOs may have leveraged larger private sums. Some farmers covered by the Chaitanya project indicated that they could not afford to contribute 50 percent of the cost of investment, but other

Figure 8.2 SWC investment by plot slope and source of finance

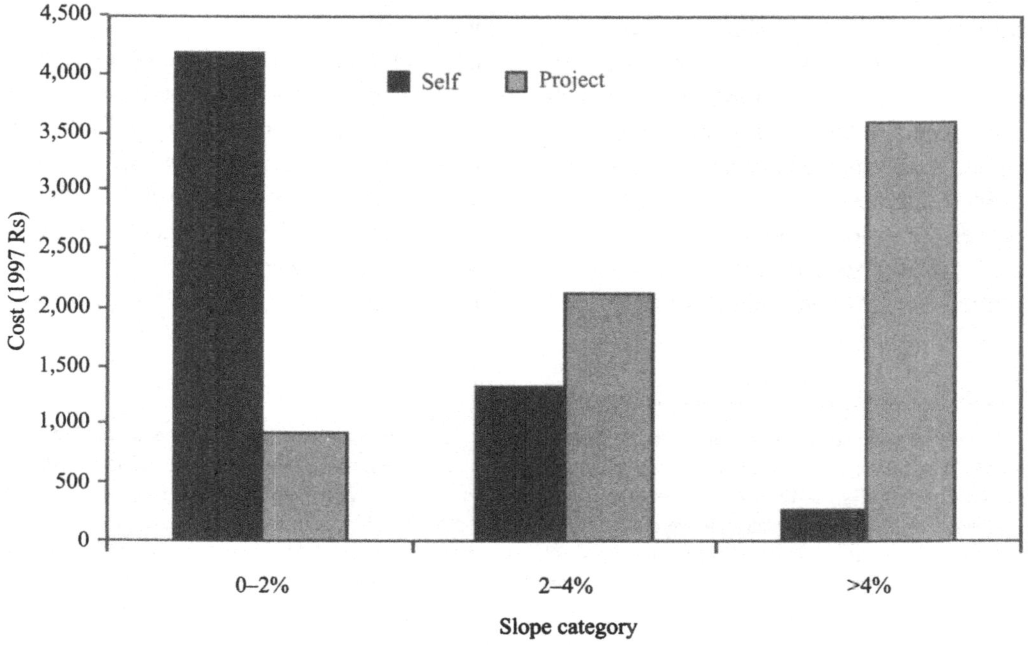

farmers did invest large sums of their own money. Perhaps Chaitanya could have had a stronger impact if it helped farmers gain access to credit to pay the matching cost.[32] In Figure 8.1, very few respondents in NGO villages used credit for land improvement investments on rainfed plots.

Use of Credit for Land Improvement Investments

The average amount of credit for soil conservation investments was only around Rs 360 out of an average total investment of nearly Rs 4,500 per hectare. Examining the sources of the small amount of credit that is used leads to an even more striking finding. About 50 percent was borrowed from moneylenders, nearly 50 percent more was borrowed from relatives and friends, and a trace amount—Rs 14 per respondent—came from informal credit groups. Not one farmer out of 239 in the survey borrowed even a single rupee from a bank for investments in land improvement on rainfed plots during the 10-year period under investigation. Irrigated plots, by contrast, received an average of more than Rs 23,300 credit, with almost Rs 9,000 coming from banks. Of this amount, about Rs 4,400 was for nonirrigation investments like bunds and leveling, with an average of Rs 900 coming from banks. This is consistent with the findings of Kerr and Sanghi (1992) that formal credit was not even available for such investments. Sometimes bank credit may be tied to special watershed projects, so that farmers can borrow to invest in certain approved technologies such as contour bunds. But typically such credit is useless, since farmers are not interested in the approved approaches. Farmers have their own practices, but banks do not recognize them and thus do not make loans available.

It is difficult to infer from the data presented here whether making bank credit more available to farmers would help stimulate improvement of rainfed lands. The problem is that most farmers might not want to borrow funds for rainfed plots even if they could, particularly for sloped, erosion-prone lands. If credit were made available in combination with subsidies, however, farmers might respond favorably. In fact, the experience of MYRADA and Outreach, two NGOs in the southern Indian state of Karnataka, indicates that this may be true to a limited extent (Kolavalli 1998; Fernandez 1998; Mascarenhas 1998). The key features of an approach that combines credit and subsidies would be, first, that credit must not be tied to specific technologies that farmers may not be interested in, and second, that subsidies must be low enough that farmers have to invest significant amounts of their own time or money. This is necessary to ensure that they will maintain investments.

Maintenance of Soil and Water Conservation Assets

If watershed agencies succeed in stimulating investment in soil conservation on sloping land prone to erosion, the next step is to encourage farmers to maintain the assets created by those investments. Table 8.3, which shows the percentage of soil and water measures that were well maintained on rainfed plots, indicates that investments with no subsidy were almost uniformly well maintained. Only 2 out of 82 were not. At

[32]Regression analysis presented later in this chapter suggests that land improvement investments on rainfed plots yielded low returns: Rs 1,000 worth of pre-1987 soil and water conservation investments resulted in an average increase of only Rs 22–42 in annual net returns to cultivation. This low range implies that farmers might not take advantage of credit even if it were available. On the other hand, weaknesses in the analysis, discussed below, mean that it is important not to take the specific numbers too literally. Also, the reported low returns are based on the combination of self-financed and project-financed investments. Many of the project-financed investments were unwanted by farmers; returns might be substantially higher for those that were not subsidized.

subsidies of Rs 2,500 or less, the overall maintenance level slipped to 84 percent, and at subsidies of more than Rs 2,500, it fell to 64 percent. Overall, 74 percent of subsidized investments were well maintained. (Investments on irrigated plots were almost always well maintained regardless of the level of subsidy, so they are not discussed here.)

The pattern holds when the data are examined separately by project category (Table 8.3). One noticeable feature is that the NGO and NGO/government collaborative projects have higher maintenance rates than government projects—100 percent for smaller subsidies and 79 percent for higher subsidies. On the one hand, the better performance of NGO and NGO/government collaborative projects, compared with government projects, suggests that a willingness to listen to what farmers actually want pays off. On the other hand, it might also reflect the fact that some of the NGO investments were on flat plots and thus easier to maintain. However, an examination of maintenance levels by both slope and project category, selecting only those plots with subsidized investments, showed that the good maintenance of NGO-supported investments was not limited to plots with no slope.

With subsidies exceeding Rs 2,500, the maintenance rate under the NGO and NGO/government collaborative projects declined to only 79 percent. NGOs invested in response to farmers' demands, and where subsidies were very high (100 percent for some projects), labor-intensive investments often provided employment for the farmer. Farmers may have accepted large investments that they did not intend to maintain, in order to obtain employment. As of 2001, subsidy policies remain unchanged for nearly all projects, so these dangers are still present.

An important question for policymakers is whether the high subsidy outlays are justified by the performance of the subsidized land improvements. This study is not able to address the effect of land improvement investments on production and conservation, but maintenance levels provide some information about performance. With overall maintenance levels of 74 percent (only

Table 8.3 Percentage of soil and water conservation investments that were well maintained, by project category and subsidy level, rainfed plots

Project category	No subsidy		Less than Rs 2500 subsidy[a]		More than Rs 2500 subsidy[a]	
	Total number	% maintained	Total number	% maintained	Total number	% maintained
Government[b]	31	97	35	77	26	58
NGO andNGO/government[c]	25	96	14	100	14	79
No project	26	100	0	...	4	50
Total (all categories)	82	98	49	84	44	64

Source: Authors' 1997 survey data.

Notes: This covers investments made from 1987 to 1997. See Chapter 2 or the glossary at the end of the report for full names and descriptions of the projects.

[a]Real value in 1997 rupees.

[b]This table combines all government projects into one category. These include the NWDPRA, World Bank, ICAR, Jal Sandharan, and DPAP.

[c]NGO projects and NGO/government collaborative projects are combined into one category for this table.

69 percent for government projects), highly subsidized investments covered in this study were not likely to be cost effective. A stronger commitment to cost sharing would help ensure that farmers only accepted land improvement measures that they truly want. Alternatively, if high subsidies are retained with the objective of providing employment benefits, creative mechanisms must be developed to ensure that farmers really want the resulting conservation measures.

Net Returns to Cultivation

Net returns to cultivation are an obvious plot-level indicator of agricultural productivity. Data are available for gross returns per hectare and both cash costs and imputed costs of household resources and labor. Investigators collected this information in interviews with farmers for the crops they grew in the year immediately prior to the interview. Unfortunately, no baseline data are available for this indicator, so the analysis is purely cross-sectional.

The analysis covers only rainfed plots since irrigation dwarfs other factors in determining net returns. The mean annual net return per hectare was Rs 30,589 on irrigated plots but only Rs 2,989 on rainfed plots. Omitting irrigated plots leaves a sample of 246 plots: 140 in Andhra Pradesh and 106 in Maharashtra.

The figures for mean net returns per hectare show far higher returns in Andhra Pradesh than Maharashtra, and they show generally higher figures for plots under projects with an NGO component than for plots under government projects or nonproject villages (Table 8.4). However, many other factors affect net returns per hectare, so it is important to analyze their determinants using multiple regression.

Explanatory variables for the plot-level analysis of returns to hectare include plot, household, village, and project characteristics. Values for 1997 are used for variables that change over time, since cultivation took place in 1997. The plot characteristics include area, land capability classification (which incorporates both slope and soil fertility), the rank of the plot within the farmer's overall holding, the number of seasons the plot is cultivated each year, the present value of land improvement investment it received, both during the 1987–97 period and prior to 1987. Household characteristics include the farmer's total landholding size, percentage of income that comes from off-farm sources, and the number of household workers. Village-level characteristics include the type of road connecting the village and the distance to the nearest market. Project characteristics include the predicted probability that the project operates in the village, and the project's expenditure per hectare. As in the village-level regressions, the model is specified with expenditure and project category expressed separately and interacting.

The model is estimated separately for each state, for three reasons. The first is that project activities and objectives vary by state, and the second is that conditions in the two states vary significantly. Government policies affecting agriculture do not differ greatly between the two states, but there may be other state-level differences

Table 8.4 Returns to cultivation on rainfed plots

	Average net returns per hectare (Rs/year)	
Project category[a]	Maharashtra	Andhra Pradesh
All plots	1,762	3,918
NWDPRA	712	4,133
DPAP and Jal Sandharan	505	3,849
NGO	2,935	4,542
AGY	2,255	n.a.
World Bank or ICAR	n.a.	3,662
No project	1,565	3,492

Source: Authors' 1997 survey data.

Notes: N.a. indicates that data were not available. See Chapter 2 or the glossary at the end of the report for full names and descriptions of the projects.

[a]Differences across project categories are not statistically significant. Maharashtra: $F = 1.35$, 4 df, $p < .26$; Andhra Pradesh: $F = 0.15$, 4 df, $p < .97$.

that are not accounted for in this model.[33] The third reason is that the instrumental variables model cannot be used for Andhra Pradesh because it is impossible to obtain predicted values of the project categories for a sample of only 16 villages. Predicted values obtained from a multinomial logit model for all 86 villages yielded highly inaccurate predictions for Andhra Pradesh because the 70 villages from Maharashtra dominated the probabilities. As a result, the instrumental variables model is run only for Maharashtra, and the findings without using predicted values are also presented.

The results of the econometric analysis, presented in Table 8.5, indicate that for Maharashtra, the instrumental variables model has very low explanatory power with an R^2 value of only 0.20. The model as a whole is not even statistically significant (although, with actual project categories, it is significant). The two models share the same significant explanatory variables. In both cases NGOs are associated with higher net returns per hectare, previous investment in soil conservation has a small positive impact on net returns, off-farm income is negatively associated with net returns, and net returns fall as the distance to the nearest market rises. One important difference in the results of the two models is the very high coefficient of NGO projects in the model with predicted project categories. It suggests that Rs 1,000 per hectare of investment in the village by an NGO project leads to an increase of Rs 1,047 in net returns per hectare of rainfed cultivation. (This is in addition to all other benefits of the project, such as increased irrigation.) This number seems impossibly high, especially since these projects do not focus particularly on raising rainfed agricultural production. In the model with actual project categories, Rs 1,000 of investment leads to an increase in net returns per hectare of Rs 447. Discussions with NGO project staff revealed that these projects place a premium on helping villagers gain access to government services and helping them identify marketing opportunities. These two activities could have a strong positive effect on crop income even if the projects do not engage in technical assistance, especially in the remote, underserved areas where many NGOs work. However, the numbers still seem high and may result from some preexisting condition in the villages for which the analysis could not account.

As mentioned earlier, the Andhra Pradesh model is run only with actual project values. In this model the number of seasons cultivated per year, the presence of high quality (class II) land, and the presence of an NGO show a positive, significant association with higher net returns. The presence of the NWDPRA has a negative, significant association with net returns. However, the results of this regression should not be taken too seriously because there is no correction for endogenous program placement. The negative coefficient for the NWDPRA may indicate nothing more than that the program worked in villages where productivity was low before the project. Another problem with this model is that the dummy variable for connection to a paved road had to be dropped because it was highly correlated (0.74) to the NWDPRA project category. This could introduce omitted variable bias. In summary, the findings in Table 8.5 probably do not warrant any strong conclusions given the model's weak predictive power and the inability to correct for endogenous program placement in Andhra Pradesh.

[33]When the analysis was conducted for both states together, a state-level dummy variable had to be dropped due to high correlation with other explanatory variables.

Table 8.5 Determinants of farmers' net returns to cultivation (ordinary least squares regression)

	Maharashtra[a]		Andhra Pradesh[a]
Variable	Model 1[b]	Model 2[c]	Model 2[c]
Plot area (hectares)	-411	-382	842
	(372)	(386)	(716)
Land capability class II (dummy)	1,377	1,375	2,370
	(929)	(919)	(931)
Number of seasons cultivated per year	1,771	1,813	7,728
	(1,223)	(1,201)	(2,841)**
Value of pre-1987 land improvements	22	22	42
(thousands of Rs/hectare)	(10)*	(10)*	(77)
Value of post-1987 land improvements	-10	-9	49
(thousands of Rs/hectare)	(24)	(24)	(67)
Farmer's total landholding (hectares)	7	3	3
	(19)	(17)	(17)
% of farmer's income from off-farm	-36	-37	0
	(17)*	(16)*	(17)
Number of workers in farm household	-13	-12	-70
	(164)	(166)	(43)
Paved road in 1997 (dummy)	-202	-244	n.a.
	(450)	(460)	
Distance to nearest market (kilometers)	-126	-138	-30
	(52)*	(47)*	(28)
Mean village expenditure/hectare (thousands of Rs)			
Project operated by:			
NWDPRA	80	112	-238
	(144)	(67)	(115)*
DPAP and Jal Sandharan	-66	3	260
	(215)	(166)	(200)
NGO	1,047	447	380
	(214)***	(198)*	(168)**
AGY or IGWDP	95	21	n.a.
	(326)	(211)	
World Bank or ICAR	n.a.	n.a..	-179
			(224)

Source: Authors' 1997 survey data.

Notes: Coefficients and standard errors are adjusted to account for sampling weights, clustering, and stratification. In model 1, predicted values based on the multinomial logit regression in Table 5.2 are used for the four project category variables in Maharashtra. In model 2 actual project dummy variables are used. Standard errors are not adjusted for use of predicted values. N.a indicates not available. See Chapter 2 or the glossary at the end of the report for full names and descriptions of the project.

[a]Sample sizes: Andhra Pradesh, 140; Maharashtra, 106.

[b]Model 1: Maharashtra: $F(7,2) = 5.79$, $p < .02$; $R^2 = 0.20$

[c]Model 2: Maharashtra: $F(7,2) = 63.00$, $p < .16$; $R^2 = 0.20$; Andhra Pradesh: $F(10,2) = 47.5$, $p < .21$; $R^2 = 0.33$.

*Statistically significant at 10 percent.

**Statistically significant at 5 percent.

***Statistically significant at 1 percent.

CHAPTER 9

Conclusions

This chapter returns to the three research questions asked at the beginning of this report: (1) What projects are most successful in promoting the objectives of raising agricultural productivity, improving natural resource management, and reducing poverty? (2) What approaches enable them to succeed? (3) What nonproject factors also contribute to achieving these objectives? The findings of the empirical study in Maharashtra and Andhra Pradesh lend support to the hypothesis that participatory projects perform better than their more technocratic, top-down counterparts, and that a combination of participation and sound technical input is likely to perform the best of all. Evidence about the role of economic conditions and infrastructure is more limited. This chapter summarizes these findings and offers suggestions for improving the impact of watershed projects and other development efforts in the future.

Evidence of Project Performance

Participatory Projects Perform Best

Concerning the management of common lands, projects in Maharashtra taking a more participatory approach—the AGY, IGWDP, and NGOs—performed better according to several indicators. The AGY and IGWDP were much more successful in introducing social fencing institutions, whereas villages under other projects differed little from nonproject villages. All of the projects appear to have contributed to reducing erosion in the main drainage line, but again, the AGY and IGWDP projects performed the best in this regard, followed by the NGOs and finally by the government projects. On uncultivated common lands, most of the projects helped reduce soil erosion below what was found in nonproject villages. Villages under NGO projects had the least erosion, followed by the AGY and IGWDP and then the DPAP. But the improved condition of common lands on these projects appears to have been at the expense of access to products such as fuel and fodder from the commons. Respondents in the AGY and IGWDP villages indicated that they had suffered from reduced access to fuel and fodder from common lands more than respondents under other projects.

The quantitative analysis did not yield strong conclusions about projects' efforts to develop irrigation. Available data could not capture the complexity of the fractured rock aquifers in the study villages. Changes in irrigated area during the project period could be the result of many factors for which data were unavailable. Nevertheless, in discussions with project beneficiaries those under the AGY and IGWDP were most likely to report that projects had helped raise the level of water in their wells. At the same time, none of the projects seem to have done much to assist farmers without irrigation or to help landless people gain access to the additional water generated through project efforts. Project officials under the IGWDP report that wages and days of employment for landless workers have risen in some villages as a result of the expansion of irrigated area (WOTR 1999). Based on the data used in this study, it was not possible to distinguish between short-term employment under the project and a long-term rise in labor demand.

On cultivated lands, the study focused on project efforts to raise productivity on rainfed plots since farmers already managed irrigated plots quite productively. Andhra Pradesh projects focused largely on developing rainfed agriculture, whereas those in Maharashtra focused more on developing irrigation. Many projects in Andhra Pradesh aimed to introduce conservation-oriented agronomic practices. All of the projects appear to have been successful in promoting cultivation across the slope. They also promoted investment in soil conservation structures such as bunds and terraces. Farmers under the NGO, AGY, and IGWDP projects invested more in soil conservation than those under other projects. Also, these projects were more effective than the purely government projects in using their own funds to leverage farmers' investment of private funds. Long-term maintenance of conservation structures was higher where farmers invested a higher proportion of their own funds. Also, for a given level of project subsidy, farmers did a better job of maintaining conservation investments on their land under the AGY, IGWDP, and NGOs than under the government projects. Finally, rainfed plots under NGOs enjoyed higher net returns per hectare than those under government projects, although the analysis on which this finding is based contained weaknesses.

Factors that Enable Participatory Projects to Perform Better

What factors enabled the more participatory projects to perform better? In answering this question it may be worth reiterating some of the characteristics of semi-arid areas that distinguish them from irrigated lands and the most favorable rainfed areas. In irrigated areas, transferring Green Revolution technology was relatively simple because improved seeds and other inputs were well suited to millions of farms covering huge areas. The new technology was so profitable, with relatively little risk, that farmers were willing to abandon traditional farming systems in favor of new approaches. In less favorable rainfed areas, on the other hand, the success of technical interventions often depends on location-specific biophysical and socioeconomic conditions and requires collective action by local people. Farmers pursue complex strategies for producing food and earning their livelihoods. New agricultural technologies usually incur opportunity costs by competing with one or more of the many components of the farm household economy, effectively reducing the net benefits of project interventions (Walker and Ryan 1990). Early watershed projects introduced technologies for conservation and production without any input from farmers, all on the basis of trials in experiment stations far from the villages and devoid of socioeconomic constraints. The lack of sustained maintenance or adoption under these circumstances is not surprising given the difficult conditions prevailing in many rainfed areas.

This background helps explain why people's participation is the key feature of the best watershed projects. All projects covered by this study claimed to take a participatory approach, but clearly the term "participation" meant different things in each case. In the most innovative and successful NGO projects, participation meant that local people were full partners in the watershed development program, with both the authority to determine how the project would proceeds and the responsibility to help plan, implement, and pay for it. In government programs, on the other hand, "participation" meant convincing local people to go along with the predetermined project design. The findings of this study suggest that full participation is critical to project success, and this should not be surprising given the special characteristics of rainfed areas.

Some specific characteristics of participatory projects are as follows:

They devote time and resources to social organization. The best projects included in this study employed staff trained in social organization and devoted substantial time to facilitate collective action prior to implementing watershed works.[34] On average more than 40 percent of the staff in the AGY, IGWDP, and NGOs were trained in social organization; no one working in the government projects had such training. The projects with an NGO component devoted at least a year to organizing people prior to making watershed investments, whereas the government-implemented projects never devoted more than a few weeks. These projects invested in watershed works only after villagers proved they could work collectively; this helped avoid superficial social organization that would probably not be sustained after project funds and staff were withdrawn.

They build each group's interests into the project. The best NGO projects recognized that rural communities are heterogeneous, composed of social groups with diverse, sometimes competing interests. These groups may include people of different religions, caste, landholding status, occupation, gender, and so forth. Some groups are always more politically powerful than others; the less powerful may have little or no say in decisions that affect their well-being. Watershed development can skew benefits in favor of wealthy people and impose burdens on poor people. Accordingly, some NGOs in Andhra Pradesh organized communities for watershed development by working separately with each interest group they could identify. They helped each group become organized and then mediated negotiation between groups, ultimately brokering a watershed development approach in which every interest group stood to gain from overall project success. For example, they encouraged participants to share benefits; they closed common lands on a rotational basis, ensuring that some common lands remained available; and they made sure that the poorest people gained employment benefits for as long as the common lands were closed.

Some participatory projects, particularly some NGOs in Maharashtra, devoted a great deal of effort to social organization but were less careful to address the interests of each social group. In particular, project plans were approved not on the basis of consensus among interest groups but by a simple vote requiring a majority of about 70 percent, depending on the project. This approach was easy to implement in Maharashtra with its relatively homogeneous social structure, but often it meant that the landless minority had no say in designing the project. As discussed in Chapter 6, shepherds typically had no say in project plans that removed their access to traditional grazing grounds. In some villages the shepherds ignored the grazing bans, undermining the project, while in other villages the grazing ban was enforced and the shepherds suffered. In such cases equity and productivity objectives were in conflict.

They work with farmers to design interventions and select technologies. In participatory projects studied in this research, project staff worked closely with farmers to design project interventions and select technologies to be used. This was critically important to ensure that beneficiaries truly wanted what the project had to offer. This approach required relaxing the strict

[34]The need to employ staff with social skills is not unique to India or to developing countries. The Landcare movement in Australia found that recruiting staff members with social skills contributed to improved performance there as well (Campbell 1994).

orientation toward achieving physical targets that most government projects pursued. And it also meant that local people helped finance the costs of investment.

As shown in Chapter 8, greater flexibility in choosing technology resulted in superior performance in maintaining soil conservation investments under participatory projects. Rainfed plots under these projects also realized higher net returns to cultivation. Some projects, particularly the NWDPRA and World Bank projects, still limited the farmers' role in choosing their own technologies, and maintenance of investments made under these projects was poor. But most projects with an NGO component took a much more flexible approach and had better results to show for it.

They choose the village, not the watershed, as the unit of implementation. Since successful watershed management depends on organizing communities to work together, the best projects used the village as the primary project unit rather than the watershed, which would be the logical unit in a purely technical program. They reconciled the village-based approach with the watershed orientation of the technical plan by breaking the watershed into subunits treated separately within each village. In short, they managed a watershed by assembling a set of small-scale plans, each of which made sense at the local level, and gradually building up to a larger scale. More technocratic projects, on the other hand, began with a master plan for a larger watershed and tried to make local units conform to it. Given the complexity of rainfed agriculture in the semi-arid tropics, poor performance under this approach should not be surprising.

They screen villages for enabling conditions. Before deciding where to implement watershed development, some of the best programs screened villages to ensure that they possessed geographic and social conditions conducive to successful watershed development. Such an approach makes sense where budgets are limited and not every village can have a watershed project.

Favorable social conditions are particularly important given the extent to which participatory approaches rely on project participants to help manage projects and make them successful. Also, one might argue that how the NGOs and NGO/government collaborative projects screen villages for their work is one of the most important determinants of these projects' success. The AGY and IGWDP in Maharashtra, which worked only in villages willing to practice *shramdan*, did the best job of screening villages for favorable social conditions. *Shramdan* is a good indicator of capability to undertake collective action, which can contribute to watershed project success. These same projects, as well as some NGO projects in Andhra Pradesh, made no investments until the villagers had demonstrated that they could successfully control grazing on common lands. Also, in Maharashtra NGO and Jal Sandharan projects favored villages that had no common land, thus eliminating an important source of conflict in designing and implementing a watershed plan. Details of these screening approaches are provided in Chapters 2 and 5.

It is important to note, of course, that there is no single critical factor that should be used to screen villages for project participation. Critical social organization skills, and indicators of their presence, may vary by location. For example, projects in Maharashtra selected *shramdan* as an important prerequisite, but projects in other places with different customs and traditions may find that other indicators are more important.

Regarding geographic conditions, two of the most important are the relationship between village and watershed boundaries and the opportunities for water harvesting. The latter is of course relevant in Maharashtra, where water harvesting is the major project objective. Relative uniformity between watershed and village boundaries facilitates planning and administration. Selecting watersheds that fall within village boundaries is a good idea given that watershed budgets are not unlimited.

They coordinate their work. Two kinds of organizational coordination appear to be important in watershed development. First, NGOs and government agencies have complementary strengths and can benefit from collaboration. The success of the AGY and IGWDP, two projects in which NGOs and government entities collaborated at every step of the project, demonstrates that this is so. It is important to contrast this with the approach to NGO collaboration promoted by other projects, including the Jal Sandharan and NWDPRA. They invited NGOs to work for a few weeks on social organization efforts, but this was seen as distinct from other project efforts.

Second, government watershed development efforts in India were famous for bureaucratic delays and turf wars that arose because watershed activities fell under the domains of numerous departments. Overcoming this problem is critical to raising the quality of work. Farrington and Lobo (1997) discussed the intricate approaches taken by the Indo-German Project to iron out interdepartmental administrative complications. The Jal Sandharan project, on the other hand, appeared to suffer from continued lack of coordination among departments (Pangare and Gondhalekar 1998). This difference may help explain the better performance of the AGY and IGWDP, compared with the Jal Sandharan.

Table 9.1 Priorities for developing the village (% of respondents)

Priority	Maharashtra	Andhra Pradesh
Improved medical facilities	38	64
Roads	37	37
Latrines	10	37
Drinking water	35	15
Irrigation	22	25
Improved bus service	8	26
Better electricity	10	20
Better educational facilities	18	9
Improved housing	4	17
Credit/bank	3	7
Watershed development	9	1
Veterinary service	8	2

Source: Authors' 1997 survey data.

Notes: Respondents listed multiple priorities. Other priorities (listed in descending order of frequency) were employment, dairy or milk collection center, telephone service including STD (long distance), community hall and equipment for it, government shop, ban on alcohol, vocational training, land for landless, fruit trees, horticulture, tree plantation, improved seeds and fertilizer, ban on dowry, community tractor, grain storage facility, weekly market, petrol pump, and post office. Large landholders were more interested in irrigation, watershed works, and credit; landless were more interested in housing, electricity, and latrines. No patterns were observed across project categories.

The Role of Infrastructure

Analysis presented in Chapters 6–8 gives weak support to the notion that improvements in performance in agricultural production between the pre- and post-project periods were greatest in villages with improvements in infrastructure. Erosion on common lands was lower in villages with higher population density, and net returns to cultivation fell as the distance to the nearest market increased. Stronger association might exist, but the econometric analysis suffered from the fact that changes in various types of infrastructure were found only in a small number of villages, so the sample may have been too small and the time frame too short to capture the effect. Also, analysis at the district level by Fan and Hazell (1999) clearly suggests that improved infrastructure raises agricultural productivity. This would suggest that the growing interest in India in an approach dubbed "watershed plus," in which watershed and infrastructure investments are designed to complement each other, has merit.

Another reason to believe that infrastructure is important is that respondents consistently listed various forms of infrastructural improvements as their top priority for developing their village. Many respondents made multiple suggestions (listed in Table 9.1).

As is the case with much of the data collected for this study, responses from the two states overlap but have some significant differences. In Maharashtra, the three most commonly listed priorities were improved medical facilities, better roads, and better drinking water supply, followed by increased irrigation and improved educational facilities. In Andhra Pradesh, improved medical facilities were mentioned most often by far, followed by better roads, latrines, irrigation, and better bus service. Table 9.1 shows other priorities also mentioned, including several that appeared too infrequently to warrant inclusion in the main body of the table.

While there were no significant differences across project categories, there were differences across landholding categories; larger landowners tended to be more interested in irrigation, watershed development and credit, while landless people were more interested in improved housing, electricity, and latrines (not shown in the table). Almost all the respondents who cited watershed development as a priority lived in Maharashtra, probably because they equated watershed development with irrigation development.

Infrastructure development is important regardless of the extent of people's participation, but there is also a role for participation in infrastructural improvement. In short, people should have a say in what kinds of infrastructure investments are made; this is part of the idea behind the 1992 *Panchayat Raj* legislation that increased the power of local governments. A further distinction is that people should also be able to choose between watershed and infrastructure investments. In a truly participatory environment in which villagers are equal partners, they should be able to determine whether scarce investment funds should be devoted to watershed development, infrastructure development, or both. It is easy to imagine that some villages must be in greater need of improved infrastructure than watershed development, so there should be flexibility to make this judgment. This is especially so given the small impact of the large amount of funds devoted to watershed development in the past.

Implications for Common Guidelines

The good performance of the more participatory projects bodes well for projects under the Ministry of Rural Development's 1994 guidelines, which the Ministry of Agriculture recently adopted as well. These guidelines are modeled after the approach taken by some of the better NGOs, and many of the projects under the new guidelines are implemented by NGOs. For example, the new guidelines allow implementing agencies to devote up to a year to social organization, they seek to promote equity, they give local people the flexibility to choose the technology they prefer, and they work within the village as the unit of operation. Accordingly, there is reason to be optimistic that projects under the new guidelines will perform better than the old, top-down and technocratic government projects examined in this study.

In fact, some early studies suggest that the projects under the new Ministry of Rural Development guidelines are performing well. Hanumantha Rao (2000) reviewed some of these studies and found increases in crop yields, in the area under cultivation in the post-rainy season, and in the availability of fodder, employment, and milk production, and a reduction in migration. Two studies reviewed were based on detailed surveys in Gujarat (Anil Shah 2000; Shah and Memon 1999); they found impressive productivity gains even though the projects had operated for only four to five years. Other studies in Andhra Pradesh were less rigorous but also suggested positive results.

At the same time, these projects appear to face the same difficulties in achieving equitable outcomes as in those examined in this research. Amita Shah (2001) analyzed the same data as in Shah and Memon (1999) and found that benefits were heavily skewed toward a minority of wealthy families with irrigated land. Anil Shah (1999) found that issues of interest to women were relatively neglected. The 1994 Ministry of Rural Development guidelines said relatively little about equity (India, Ministry of Rural Development 1994a), but project officials soon saw the need to make special efforts to ensure that the poorest people share the gains from watershed development.[35] In fact, the common guidelines agreed upon by the Ministry of Agriculture and the Ministry of Rural Development in 2000 state explicitly that ways need to be found to share the benefits of watershed development (such as water and biomass) with landless and near landless people (India, Ministry of Agriculture and Cooperation 2000). Unfortunately the guidelines do not suggest any specific approaches to achieve this, but equity is clearly becoming a high priority for the government. To be fair, to date few if any other projects have been able to easily implement ways to share project benefits. This is an important area for exploration and experimentation.

Another important concern about the new generation of participatory government projects is that they are trying to expand projects to a huge scale, using an approach that so far has only proven successful at a small scale. The NGO/government collaborative projects analyzed in this study included at most only a few hundred villages, whereas the government projects aim to operate in thousands of villages. The NGO/government collaborative projects performed well in this study, but it is important to recognize that they benefited from favorable treatment that cannot be extended to a large scale. For example, all of their villages had been part of previous watershed projects (as had almost all other projects in Maharashtra). In all of the IGWDP's sites covered under this study, an experienced NGO had already been active in the village for several years. The AGY, meanwhile, was a high profile project subject to relatively frequent visits from high-ranking government officials. As a result, project staff may have worked particularly hard, and development funds for all kinds of activities were allocated on a priority basis. Such special treatment will not be possible as these projects continue to expand, so it is premature to draw conclusions about the potential for scaling up based on the findings presented here. Additionally, depending on NGOs to implement projects may work well on a small scale, but there are not enough high-quality NGOs to do so on the vast scale of the Ministry of Rural Development's watershed program. Building the capacity of NGOs to manage participatory watershed projects is an important task. Finally, the same technocratic project officials who oversaw top-down approaches to watershed development for many years will play major roles in efforts to increase the level of local participation in the new government projects. Expecting them to rapidly transform their mindset from supervisor to facilitator is unrealistic; it will take time and encouragement. These comments are not meant to detract from the good performance of the IGWDP and the AGY, but rather to sound a note of healthy caution.

Additional Issues for the Future

The previous section has suggested a number of approaches that contribute to better project performance, and all projects should strive to pursue them. Two broader issues important for better allocation of resources to watershed development are the need for

[35] Personal communication with S. P. Tucker, former director of rural development, Andhra Pradesh, in October 1996 and with N. K. Sanghi, National Institute of Extension Management, Hyderabad, in April 2001.

caution in applying subsidies to watershed investments and the need for better monitoring and evaluation.

Subsidies for Soil and Water Conservation

Subsidies for soil and water conservation present complicated problems that all watershed projects must try to solve. On the one hand, soil and water conservation in upper catchments generates water-harvesting benefits that accrue to downstream farmers, implying a positive externality for which farmers should be compensated. On the other hand, this research shows a clear negative correlation between the level of subsidy and the likelihood that a farmer will maintain a soil conservation investment in good condition. The reason for this is that many projects either impose on farmers soil conservation structures that they do not want, or they offer such high payment—sometimes with project wages that exceed market wages—that farmers accept them voluntarily even if they do not really want them. In either case, the farmer may have an incentive to remove the structure after the project staff leave.

Obviously, part of the solution is to avoid imposing structures on farmers without their consent. Another part of the solution is to learn from farmers what kinds of conservation practices they really do want for their own merits, not simply because their construction is more labor-intensive. Both of these problems may diminish as more participatory approaches become institutionalized, the former because projects will become more farmer-friendly and the latter because farmers will lose their preconceived expectation that projects will bring no benefits except short-term employment and assorted giveaways. However, as long as subsidies approach 100 percent, there is a danger that project officials will be unable to know whether a farmer is responding to a promising investment opportunity or just to the promise of employment.

The issue of compensation to upstream farmers for downstream water-harvesting benefits is integrally linked to that of equitable distribution of project benefits. The fundamental social problem of watershed development is that it often distributes benefits and costs unevenly, making it a likely source of disagreement and conflict. On the bright side, in most of the semi-arid areas of India watershed externalities are highly localized, operating within villages or between adjacent villages rather than between a large, dispersed group of small land users whose actions lead to siltation of a distant water body. As a result, devising solutions to the problem may be simpler. One obvious approach is to develop mechanisms to share the increased water and biomass benefits that watershed development generates. The new common guidelines between the Ministry of Agriculture and the Ministry of Rural Development endorse this idea, but they do not offer suggestions on how to put it in practice. If this could be achieved, however, it would create an unambiguous incentive for upstream farmers to install soil and water conservation practices that protect downstream water sources.

Two of India's early success stories in watershed development, the Sukhomajri and Pani Panchayat projects, operated on the principle of full and equal water rights for all households, both landed and landless. Watershed development under these projects continues to operate successfully more than 25 years after it was introduced. The universal benefits that it generates are undoubtedly an important reason for their success.

A few contemporary projects have succeeded in promoting partial sharing of additional natural resource benefits generated by projects, but the overall record is weak. With groundwater the issue is particularly difficult, because Indian water law states explicitly that any landowner is entitled to water pumped from beneath his land, as long as it does not interfere with drinking water supplies (Singh 1991). As a result, project authorities can try to negotiate arrangements to share groundwater, but they cannot force landowners who dissent. Legislative reform in this area would be extremely unpopular among those who own irrigation wells, but it could provide an important means of making watershed development more sustainable.

Monitoring and Evaluation

This research suffered from a scarcity of good data on agricultural productivity and natural resource conditions, but this lack of information has other more serious implications. In particular, it means that government planners lack sufficient data to draw firm conclusions about the returns to different kinds of watershed development investments. Given the vast size of the budget for watershed projects, better information about their performance would go a long way toward more cost-effective government planning. Currently too many funds are allocated on the basis of too little information and, as the findings from this study show, the potential for waste is great.

The data shortage takes two forms: first, a lack of baseline data against which to compare current conditions, and second, a lack of monitoring data for easy assessment of current conditions.

Baseline data. Most projects collect at least a small amount of baseline data while selecting project sites and preparing work plans. In NGO projects, background data cover both agroclimatic and socioeconomic issues, while in projects managed by state-level government departments, the data are skewed toward agroclimatic factors. This reflects the technical orientation of most government watershed agencies. Government projects typically conduct detailed soil surveys and prepare detailed land use maps before commencing work. Many NGOs may collect similar data through less formal but equally detailed participatory rural appraisal exercises. In both cases, however, typically there is no systematic mechanism for storing the data and making it available for comparison at a later date. Inquiries with government offices revealed that such records are often discarded once the project work comes to a close. The reason is that for both government and nongovernment projects, baseline data are usually collected for the purpose of planning, not evaluation.

Monitoring. All government watershed projects keep detailed records of funds spent, structures built, and other physical targets, but such information reveals nothing about impact. It is purely a bureaucratic requirement to limit misuse of funds. Most NGOs also keep records of work done, and a small number of the better ones evaluate their own work. The World Bank's IWDP provides a clear example of collection of detailed monitoring and evaluation data. Tthis work is contracted to researchers at state agricultural universities who produce regular, detailed reports on the performance of technical interventions. The NWDPRA also has guidelines for monitoring and impact evaluation.

Three important problems remain, however. First, it is difficult to obtain the data that have been collected for monitoring. For example, efforts to obtain such data for the current study, sponsored by the government of India, were not successful. Second, the data are not organized in a common format across different types of projects, so they are not necessarily useful for comparison between project types. Third, the monitoring procedures under some projects, such as the IWDP and NWDPRA, fail to address socioeconomic issues or the implementation process. In the future, monitoring should address process in order to obtain a better understanding of the challenges and effects of participatory approaches.

Common monitoring and evaluation guidelines are needed. There is a strong need to develop common guidelines for collecting baseline and monitoring data. But what kind of information should be gathered and at what level? It is best to keep the data set small so collecting and maintaining it do not become a burden. It would be easy for the Ministry of Agriculture and the Ministry of Rural Development to issue guidelines for evaluating all projects within their jurisdiction, but generating common guidelines acceptable to multiple ministries and even NGOs would be more difficult. Accordingly, a high level meeting to develop a common framework for data collection should be a high priority. Such a gathering should include not only ministry officials but also representatives of NGOs and researchers in order to make sure that all parties' priorities are addressed and that a workable, usable system is developed. A tiny proportion of the vast watershed budget in each ministry could then be set aside for collecting and maintaining such data for a representative sample of all kinds of watershed projects throughout the country. A common interministerial office could be responsible for monitoring watershed projects. Arrangements could be made to gather data from all kinds of projects, including those of NGOs.

Glossary

AGY	Adarsh Gaon Yojana, or Ideal Village Scheme, operated by the state of Maharashtra in collaboration with local NGOs.
Bund	Soil conservation barrier made of earth, stone, or vegetation. Also the earthen dam that forms an irrigation tank or percolation tank.
Check dam	A small earthen or stone structure that captures runoff water in a drainage line. Like a percolation tank but smaller in scale.
COWDEP	ComprehensiveWatershed Development Program , an early watershed project in Maharashtra.
CRIDA	The Central Research Institute for Dryland Agriculture, located in Hyderabad, is part of the Indian Council of Agricultural Research, sponsored by the Central Government. Its mandate is to promote productive rainfed agriculture in areas with low rainfall.
DPAP	Drought Prone Areas Program, funded by the Indian Ministry of Rural Development.
Drainage line	A dry streambed or gully where water concentrates during heavy rainfall.
ICAR	Indian Council of Agricultural Research, a group of national research institutes
ICRISAT	International Crops Research Institute for the Semi-arid Tropics, a Future Harvest center located in India
IGWDP	Indo-German Watershed Development Project, operated by NGOs in collaboration with the government of Maharashtra.
Irrigation tank	An artificial pond that collects runoff water that is used for gravity-fed irrigation. Water is distributed by gravity through a gate in the tank bund to irrigate the fields below.
IWDP	Integrated Watershed Development Project, operated by state governments with financial support and technical assistance from the World Bank.

Jal Sandharan	The Water Conservation Department of the Government of Maharashtra.
JRY	Jawahar Rojgar Yojana is a program funded by the central government to provide wage employment for rural people.
MYRADA	An NGO located in Bangalore, Karnataka, which works in several southern states including Andhra Pradesh.
NCAP	National Centre for Agricultural Economics and Policy, an Indian government research institute in New Delhi, which helped conduct the 1997 survey for this study.
NGO	Nongovernmental organization
NGO/government	Collaboration between nongovernmental and governmental organizations
NWDPRA	National Watershed Development Project for Rainfed Areas, operated by the Indian Ministry of Agriculture.
Panchayat	Village government
Percolation tank	An artificial pond that collects runoff water, causing it to infiltrate into the ground to recharge groundwater aquifers.
Pilot Project	Refers to the World Bank Pilot Project for Watershed Development in Rainfed Areas, which operated in four Indian states from 1984 to 1991.
Sarpanch	A village's political leader.
Shramdam	Voluntary community labor that contributes to community welfare.
Social Centre	NGO working in Maharashtra, which was founded by Jesuits in 1969
Social fencing	Social mechanisms used to protect common lands
Taluka	A subdistrict administrative unit containing up to 200 villages

References

Agarwal, A., and S. Narain. 1989. *Towards green villages: A strategy for environmentally sound and participatory rural development.* New Delhi: Centre for Science and Environment.

Agresti, A., and B. Finlay. 1997. *Statistical methods for the social sciences,* 3rd ed. Upper Saddle River, N.J., USA: Prentice Hall.

Ahluwalia, M. 1997. Representing communities: The case of a community-based watershed management project in Rajasthan, India. *IDS Bulletin* 28(4): 23–34.

Bentley, W. R. 1984. *The uncultivated half of India: Problems and possible solutions.* Discussion Paper No. 12. New Delhi: Ford Foundation.

Campbell, C. A. 1994. *Landcare: Communities shaping the land and the future.* St. Leonards, N.S.W., Australia: Allen and Unwin

Chambers, R., T. Shah, and N. C. Saxena. 1989. *To the hands of the poor: Water and trees.* New Delhi: Oxford and IBH Publishing Co.

Chopra, K., G. Kadekodi, and M. N. Murty. 1990. *Participatory development, people and common property resources.* New Delhi: Sage.

Ezemenari, K., A. Rudqvist, and K. Subbarao. 1999. Impact evaluation: A note on concepts and methods. Poverty Reduction and Economic Management Network, World Bank, Washington, D.C. Photocopy.

Fan, S., and P. Hazell. 1999. Are returns to public investment lower in less-favored rural areas? An empirical analysis of India. Environment and Production Technology Division Discussion Paper 43. International Food Policy Research Institute, Washington, D.C.

Farrington, J., and C. Lobo. 1997. Scaling up participatory watershed development in India: Lessons from the Indo-German watershed development programme. Natural Resource Perspectives Discussion Paper 17. Overseas Development Institute, London.

Farrington, J., C. Turton, and A. J. James, eds. 1999. *Participatory watershed development: Challenges for the Twenty-First Century.* New Delhi: Oxford University Press.

Fernandez. A. P. 1993. *The MYRADA experience: Alternate management systems for savings and credit of the rural poor.* Bangalore: MYRADA.

_______. 1994. The *MYRADA experience: The interventions for a voluntary agency in the emergence and growth of people's institutions for sustained and equitable management of micro-watersheds.* Bangalore: MYRADA.

_______. 1998. The MYRADA *experience: Will people take loans for treatment measures on private lands in a micro watershed?* Bangalore: MYRADA.

Gadgil, M., and R. Guha. 1992. *This fissured land: An ecological history of India.* New Delhi: Oxford University Press.

Gebremedhin, B., J. Pender, and G. Tesfaye. 2000. Community natural resource management: The case of woodlots in Northern Ethiopia. Environment and Production Technology Division Discussion Paper 60. International Food Policy Research Institute, Washington, D.C.

Greene, J., and V. Caracelli. 1997. *Advances in mixed-method evaluation: The challenges and benefits of integrating diverse paradigms.* New Directions for Evaluation 74. San Francisco: Jossey-Bass.

Greene, W. 1990. *Econometric analysis.* Englewood Cliffs, N.J., USA: Prentice Hall.

Hanumantha Rao, C. H. 2000. Watershed development in India: Recent experience and emerging issues. *Economic and Political Weekly,* November 4: 3943–3947.

Hayami, Y., and V. Ruttan. 1984. Induced innovation model of agricultural development. In *Agricultural development in the Third World,* ed. C. Eicher and J. Staatz. Baltimore, Md., USA: Johns Hopkins University Press.

Hazare, A., G. Pangare, and V. Lokur. 1996. *Adarsh Gaon Yojana: Government participation in a peoples programme (Ideal Village Project of the Government of Maharashtra).* Pune, India: Hind Swaraj Trust.

ICRISAT (International Crops Research Institute for the Semi-Arid Tropics). 1996. Economics of pasture protection and development in Rajasthan: Interim report submitted to the Department of Watershed and Soil Conservation, Government of Rajasthan. Patancheru, India. Photocopy.

India, Ministry of Agriculture. 1991a. *WARASA: National Watershed Development Project for Rainfed Areas (NWDPRA) Guidelines,* 2nd ed. New Delhi: Ministry of Agriculture.

India, Ministry of Agriculture and Cooperation. 2000. Common approach for watershed development. New Delhi. Also at <http://www.nic.in/agricoop/guide.htm>.

India, Ministry of Rural Development. 1994a. Guidelines for watershed development. New Delhi.

______. 1994b. *Report of the Technical Committee on Drought-Prone Areas Programme and Desert Development Programme.* New Delhi: Ministry of Rural Development.

India, Office of the Registrar General. 1991. *Census of India 1991.*

Jalan, J., and M. Ravallion. 1998. Transfer benefits from welfare: A matching estimate for Argentina. World Bank, Washington, D.C. Photocopy.

Kerr, J., and N. K. Sanghi. 1992. *Indigenous soil and water conservation in India's semi-arid tropics.* Sustainable Agriculture Programme Gatekeeper Series Paper No. 34. London: International Institute for Environment and Development.

Kerr, J., N. K. Sanghi, and G. Sriramappa. 1996. *Subsidies in watershed projects: Distortions and opportunities.* Sustainable Agriculture Programme Gatekeeper Series Paper. London: International Institute for Environment and Development.

Kerr, J., G. Pangare, V. Lokur, and P. J. George. 1999. Effects of watershed project subsidies on soil and water conservation investments in India's semi-arid tropics. In *Using incentives in soil conservation,* ed. D. Sanders, P. Huszar, S. Sombatpanit, and T. Enters. Enfield, N.H., USA: Science Publishers.

Kolavalli, S. 1998. A review of approaches to watershed development in India. A report submitted to the National Centre for Agricultural Economics and Policy Research, Indian Council of Agricultural Research. New Delhi.

Maddala, G. S. 1983. *Limited-dependent and qualitative variables in econometrics. Cambridge* and New York: Cambridge University Press.

Manski, C. 1995. *Identification problems in the social sciences.* Cambridge, Mass.: Harvard University Press.

Mascarenhas, J. 1998. *Organisational and human resource development aspects of enhancing cooperation between people and institutions.* Outreach Series Paper 10. Bangalore, India: Outreach.

Mascarenhas, J., J. Jangal, P.D. Prem Kumar, R. Rathod, D. Naik, and Maidrappa. 1991. Community organization and participatory learning methods in the watershed context. In *Farmers' practices and soil and water conservation programs,* ed. J. Kerr. Summary proceedings of a workshop. Patancheru, India: ICRISAT.

Morduch, J. 1999. The microfinance promise. *Journal of Economic Literature* 37(4): 1569–1614.

NABARD (National Bank for Agriculture and Rural Development). 1995. *Guidelines on participation in Indo-German Watershed Development Programme.* Mumbai, India.

Pangare, G., and S. Gondhalekar. 1998. Evolution of the watershed development program in Maharashtra: A holistic view. Paper presented at the National Workshop on Watershed Approaches for Managing Degraded Land in India: Challenges for the 21st Century, April 27–29, 1998, Department of Wastelands Development, New Delhi.

Patel-Weynand, T. 1997. Sukhomajri and Nada: Managing common property resources at the village level. In *Natural resource economics: Concepts and applications to India,* ed. J. Kerr, D. K. Marothia, K. Singh, C. Ramasamy, and W. R. Bentley. New Delhi: Oxford and IBH.

Patton, M. Q. 1997. *Utilization-focused evaluation,* 3rd ed. Thousand Oaks, Calif., USA: Sage.

Pender, J., and J. Kerr. 1998. Determinants of farmers' indigenous soil and water conservation investments in semi-arid India. *Agricultural Economics* 18(2): 113–125.

———. 1999. The effect of transferable land rights on credit, land investment and use: Evidence from south India. *Agricultural Economics* 21: 279–294.

Pender, J., and S. Scherr. 1998. Organizational development and natural resource management: Evidence from central Honduras. Environment and Production Technology Division Discussion Paper 49. International Food Policy Research Institute, Washington, D.C.

Pitt, M., and S. Khandker. 1996. *Household and intrahousehold impact of the Grameen Bank and similar targeted credit programs in Bangladesh.* World Bank Discussion Paper No. 320. Washington, D.C.: World Bank.

Pitt, M., M. Rosenzweig, and D. Gibbons. 1993. The determinants and consequences of the placement of Government Programs in Indonesia. *World Bank Economic Review* 7(3): 319–348.

RAU (Rajasthan Agricultural University). 1999. Impact evaluation report: World Bank Integrated Watershed Development Project (Plains), Rajasthan. College of Technology and Agricultural Engineering, Rajasthan Agricultural University, Udaipur.

Sanders, D., P. Huszar, S. Sombatpanit, and T. Enters, eds. 1999. *Using incentives in soil conservation.* Enfield, N.H., USA: Science Publishers.

Seva Mandir. 1999. *Annual report,* 1998–99. Udaipur, Rajasthan, India.

Shah, Amita. 2001. *Who benefits from participatory watershed development? Lessons from Gujarat, India.* Sustainable Agriculture and Rural Livelihoods Programme Gatekeeper Series No. 97. London: International Institute for Economic Development.

Shah, Amita, and G. Memon. 1999. Watershed development projects in Gujarat: A quick review. Gujarat Institute of Development Studies, Ahmedabad, India.

Shah, Anil. 2000. Eloquent silent revolution. Development Support Centre, Ahmedabad, India.

Shah, T. 1993. *Groundwater markets and irrigation development: Political economy and practical policy.* New Delhi: Oxford University Press.

Singh, C. 1991. Water rights and principles of water resources management. Indian Law Institute, New Delhi.

Singh, K. 1997. Property rights and tenures in natural resources. In *Natural resource economics: Concepts and applications to India,* ed. J. Kerr, D. K. Marothia, K Singh, C. Ramasamy, and W. R. Bentley. New Delhi: Oxford and IBH.

Stata Corporation. 1999. *User's guide and reference manual.* College Station, Tex., USA: Stata Press.

Tiffen, M., M. Mortimore, and F. Gichuki. 1994. *More people, less erosion: Environmental recovery in Kenya.* Chichester, U.K.: John Wiley and Sons.

Trochim, W. M. 2000. *The research methods knowledge base,* 2nd ed. Internet WWW page, at URL: <http://trochim.human.cornell.edu/kb/index.htm>. Updated August 2000.

Turton, C., J. Coulter, Anil Shah, and J. Farrington. 1998. *Participatory watershed development in India: Impact of the new guidelines.* London: Overseas Development Institute,.

Vedeld, T. 2000. *Participation in the Himalayan foothills: Lessons from watershed development in India.* Social Development Paper No. 38. Environmentally and Socially Sustainable Development Network, South Asia. Washington, D.C.: World Bank.

Wade, R. 1988. *Village republics: Economic conditions for collective action in South India.* Cambridge: Cambridge University Press.

Walker, T., and J. Ryan. 1990. *Village and household economies in India's semi-arid tropics.* Baltimore, Md. USA: Johns Hopkins University Press.

World Bank. 1988. *India review of rainfed agriculture and rural development.* Washington, D.C.

WOTR (Watershed Organization Trust). 2002. *Watershed Organization Trust, Ahmednagar, Maharashtra* < http://www.wotr.org> (accessed December 2002).

______. 1999. Watershed development and the landless. *The Watershed Family: Newsletter of the Indo-German Watershed Development Programme* 2 (January).